CW00505297

# Risk and Safety in Engineering Processes

By

Ivan Lucic

Cambridge
Scholars
Publishing

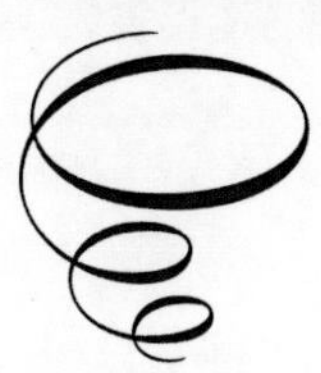

Risk and Safety in Engineering Processes

By Dr. Ivan Lucic

This book first published 2015

Cambridge Scholars Publishing

Lady Stephenson Library, Newcastle upon Tyne, NE6 2PA, UK

British Library Cataloguing in Publication Data
A catalogue record for this book is available from the British Library



ISBN (10): 1-4438-7077-3
ISBN (13): 978-1-4438-7077-1

My life and my research would be impossible without the love and support of my loving wife, Marija, and the joy and inspiration brought to me by our daughter Anka and son Lazar. I will always be grateful to my wife for her loyal support whenever and whatever I did, however foolish it may have been.

# CONTENTS

# LIST OF FIGURES

# LIST OF TABLES

# FOREWORD

It has been my pleasure to have known and worked with Ivan for the last five years. His passion for work and his work ethic towards making potentially complex and complicated procedures easier to understand, not only for professionals in the risk and safety field, but also to the lay person, is motivational.

A lot of research went into the production of this book, looking at how risk and safety was managed in the past, and where either additional clarity or changes to the processes were required. The output from this research, and this book, has been developed into the Engineering Safety and Assurance Case (ESAC) that has been used on London Underground on it's last two major projects over 8 years, the Victoria Line Upgrade programme (£1.5 billion cost) and the Sub-surface Upgrade Programme (approx £5 billion cost).

The ESAC combines all the arguments and evidence required to prove that the Engineering Safety, Reliability, Availability, Maintainability, Operability and performance of the "changed Railway" is fit for purpose and meets the client's requirements. It also ensures that the appropriate amount of analysis and work is undertaken dependent on the type of change.

To date, because of the structure of the ESAC and the training Ivan has given to project / programme staff to ensure they know the when, why and what of the process, every ESAC has been delivered on time and been accepted first time every time.

This is all credit to Ivan and the motivation he gives to his staff.

Jonathan Harding
MSc, BSc, CEng, CPhys, MInstP, MIEnvSc

# PREFACE

This book is focused on the treatment of safety risks in railways. Existing methodologies for assessment and management of the safety risk on railways are mostly empirical, and have been developed out of a need to satisfy the regulatory requirements along with in response to a number of major accidents. Almost all of these processes and methodologies have been developed in support of approvals of specific products or very simple systems, and do not add up to a holistic, coherent methodology that would be well suited for the analysis of modern, complex systems, involving many vastly different constituents (software, hardware, people, products developed in different parts of the world, etc.). The complexities of modern railway projects necessitate a new approach to risk analysis and management.

At the outset, the focus of the book is on the organisation of the existing family of system analysis methodologies into a coherent, heterogeneous methodology. An extensive review of existing methodologies and processes was undertaken, and is summarised here. Relationships between different methodologies and their properties were investigated seeking to define the rules for embedding these into a hierarchical framework and relating their emergent properties.

Four projects were utilised as case studies for the evaluation of existing methodologies, processes, and initial development. Later, this book describes the methodology adopted in support of the development of the System Safety Case and its structure.

Based on that experience and knowledge, a set of high level requirements was identified for an integrated, holistic system, safety analysis, and management process. A framework consisting of existing and novel methodologies and processes was developed and tried on a real life London Underground project. During the trial, several gaps in the process were identified and adequate new methodologies or processes were defined and implemented in order to complete the framework.

The trial was successful and the new framework, referred to as the Engineering Safety and Assurance Case Management Process, has now been implemented across the London Underground Capital Programmes Directory.

**Key words:** Risk modelling, Systems Approach, Holistic, Safety Case, Systems Assurance, Change Safety Analysis, System Safety, System Integration.

# ACKNOWLEDGMENTS

I am extremely grateful to Prof Nicos Karcanias and Prof Ali Hessami for their patience, guidance, and above all, friendship. It was them who 'tricked' me into commencing this expedition into the enchanted world of research where everything is possible. I am grateful to Nicos for his support, advice and guidance. His constructive criticism and observations focused my thoughts and directed my journey. I owe an immense debt to Ali for all the knowledge and understanding of systems and safety theory and practice, long discussions about ethics, help, support, and trust he had in me for all these years. I am grateful for my mother's persistent, subtle encouragement to complete the research and my father's nonchalant faith in me.

I am immensely thankful to friends and colleagues from my team in LU who work with me, patiently enhancing my understanding and stubbornly helping me to implement my ideas on very challenging projects. So thank you, very much, Jon Harding, Xenophon Christodoulou, Roger Short, Dr. Bruce Elliott, Ian Shannon, Dr. Lucy Regan, Peter Stanley, Rob Jones, Mukesh Sharma, Michael Mangroo, Mike Lester, Ian Innes, Paul Lawless, Tim Ballantyne, Chi Wang and Dr. Josh Ahmed.

A very special thank-you, to Ricardo Hetherington, a very good friend of mine, and the editor of the book.

# CHAPTER ONE

# INTRODUCTION

## 1.1 Problem area

In engineering problems, detailed analyses of risk and its attributes can lead to significant benefits in safety performance along with savings in time and money.

Most of the existing processes and guidelines state what needs to be a result of the safety risk analysis, broadly outlining the expectations from each of the identified risks and depicted activities (listed later). Alternative methodologies exist for some of the activities, but not for all; each one invented and used within the context of different aspects of the system's structure, behaviour and/or its emerging properties.

However, none of these amount to an integrated framework for the analysis and management of a system's safety risk .

For example, any sound analysis, including risk analysis, should be based on a series of observations and measurements. The first stage of this activity, before the hazard identification can be carried out, should be to define a system to be analysed, including the definition of the scope and context of the analysis, and the development of some form of system description. This preparation should also support the identification of the experience and expertise of the participants of the hazard identification process. Yet none of the guidelines provide any support in this area.

Furthermore, none of the processes depicted in the available literature provide any guidance in relation to the monitoring of changes to the system during its lifecycle, and the control of the impact of that change on the safety performance of the system. These are just some of the shortcomings of the existing processes and guidelines.

The scope of this research includes engineering safety analysis and management that is applicable to any industry, but with the trial implementation being specific to the railway industry. The methodology aims to support the systems safety analysis and the identification of major contributors to safety risks and benefits, whilst safety and business decision-making is supported through the evaluation of different application solutions and mitigation measures. The methodology thus supports the holistic evaluation of safety risk and alternative solutions to significantly improve safety and economic performance. Later in the book, this methodology is referred to as Engineering Safety and Assurance Process (ESAP).

## 1.2 Objectives and Aim of the book

Society generally expects a level of safety from products and services, and the current legislation (EU as well as UK) supports this view.

History demonstrates that safety failures can have significant societal costs, life or health, monetary and environmental. Again, history shows that most, if not all, accidents are avoidable.

Thus, society and legislation dictate that we all have a 'Duty of Care' to the following groups:

1 Staff and Colleagues,
2 Passengers,
3 Members of the Public, and
4 The environment.

The complexity of modern projects and products demands the system's approach to the analysis of safety as an emerging property of the system itself, for the simple reason that with increased complexity of the systems produced by human kind, our ability to comprehend the totality of the system without a structured methodical process is decreasing.

The objective of the research that preceded this book was to develop an innovative, integrated methodology in support of safety risk assessment and management for engineering problems, in particular in support of the introduction of large scale, novel and complex railway systems. However, the developed process is generic and can be used in any industry or undertaking. The aim of the research was to contribute to decision-making and management practices involving safety risk. The research was carried out in three stages:

1. Research of existing industry practices and literature;
2. Application, testing and improvement of a selection of the existing processes on four real life projects. Development of new methodologies for risk assessment as part of this work;
3. Further development of the integrated, innovative methodology, and finally testing and implementation on a real life project.

Once an outline framework had been created by utilising the existing methodologies on real life projects, the research focused on the development of methodologies in support of the activities not catered for, or not sufficiently supported, by the existing methodologies, along with integrating these new methodologies, with the existing ones, into a new holistic process.

Three high level principles of analysis and management of safety risks must be respected and supported by adequate processes, as only through adherence to these principles will one be able to:

1 Ensure completeness of analysis;
2 Build a defendable argument in support of the final results (not forgetting that the choice based on the analysis may directly influence decisions potentially affecting human lives and costing millions).

These principles are:

1 Systematic Approach to defining the problem space;
2 Holistic approach to the analysis and assessment;
3 Necessity of extensive use of Domain Specific Expertise.

After an initial literature review, and following on from the experience of working on a number of railway related safety case development projects, a number of major steps have been identified as generic safety risk analysis and management process activities (listed later). Processes and tools in support of missing, or at least insufficiently developed, stages, were invented or further developed, and following that, an integrated process which was inclusive of all these steps was developed.

In summary, the result of the research project, reported in this book, is a decision supporting methodology, that was used as part of making and managing decisions involving system safety risk, and in the development of the safety cases on one of the most complex railway projects in the UK, the upgrade of London Underground's Victoria Line (Lucic and Short, 2007) and Subsurface Lines (Metropolitan, District, Circle and Hammersmith & City and East London).

## 1.3 Structure of this book

This book is structured into 8 chapters as outlined by the use of Goal Structuring Notation (GSN) in Figure 1-1 below. As already mentioned, the author made significant use of the existing methodologies and processes, combining them with novel methodologies and processes (developed as part of this research), in order to transform them into a new general framework for safety risk analysis and management. GSN elements corresponding to novel processes and/or methodologies are indicated by red line outlines. Later on, in the introduction to each chapter, discussion of the novel use of existing processes and methods has been indicated in blue and the application of existing methods or processes within the new framework has been indicated in green.

A more detailed outline of the logical argument of this book is provided within introduction for each of chapters 3, 4, 6 and 7 using Goal Structuring Notation.

The First chapter outlines the aim, objectives and scope of the book, and presents the structure that we will follow. It also presents the more important definitions that will be used later on.

The Second chapter portrays the history of the problem area and the background to the research. It also defines the problem area and the aims and objective of research.

The Third chapter presents the findings of the literature review, as well as providing an overview of basic processes and tools for analysis, treatment and the modelling of safety risk in engineering processes.

The Fourth chapter outlines the initial applied research areas, and presents the findings of the early development of the methodology. Also described are four real life projects that have been used as a vehicle for initial development and testing.

The Fifth chapter presents a critique of existing tools for the analysis and management of safety risks and, using experience gained on the real life projects outlined in the fourth chapter, sets out the agenda for the final phase of the research.

The Sixth chapter develops the theoretical background of the basic concepts further and lays out a new system-based framework for safety risk analysis and management.

The Seventh chapter presents the challenges, results and observations of the application of the new framework on two real life projects.

The Eighth chapter concludes the findings of the research and outlines the requirements and direction of further research.

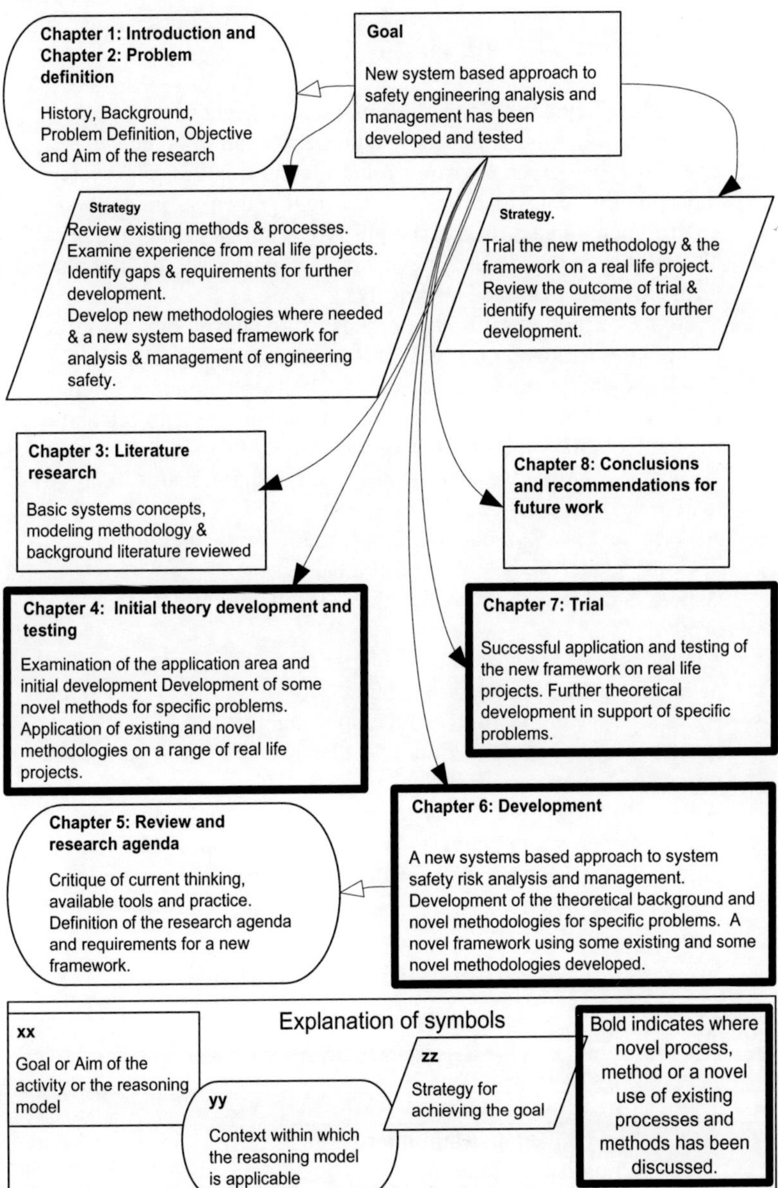


Figure 1-1: Overall Structure of the book

## 1.4 Definitions

The following are key definitions of terms used throughout the book:

1. A System is an interconnection, an organisation of objects that is embedded in a given environment (Karcanias, 2003). The system is the sum of all constituent parts working together within a given environment to achieve a given purpose within a given time period. The totality of the system is a matter of perspective. It is not a fixed term, but can be defined arbitrarily;
2. System Conceptualisation is a process of the development of the internal (constituent parts and their connections) and external (environment) system specification. A conceptual model should reflect knowledge about the application domain rather than about the implementation of the system (Milloti, 2004);
3. A Hazard is defined as an object, act or condition that is likely to lead to an accident;
4. An Accident is an unplanned, unintended event, entailing loss;
5. A Consequence is the outcome of a hazard;
6. A Loss is defined as an undesirable, detrimental effect of an accident;
7. An opposite of the hazard is an Opportunity. This is an object, act or condition likely to lead to a gain;
8. A Gain is a desirable effect of the opportunity;
9. A Risk is a forecast for a future accident of a certain severity. An opposite of risk is reward;
10. A Risk Profile is a multi-dimensional presentation of forecasts for future accidents, of certain severities, for a system. Additional dimensions introduced, may be time, space or some other relevant variable parameters.

## 1.5 Acronyms and Abbreviations

| Acronym | Definition |
|---|---|
| AC | Alternating Current |
| ACC | Area Control Centre |
| AGC | Agreement on Main International Railway Lines standard (English translation) |
| AGTC | Agreement on Important International Combined Transport Lines and Related Installations (English translation) |
| ALARP | As Low As Reasonably Practicable |

| Acronym | Definition |
|---|---|
| ALF | Algorithm File |
| ATO | Automatic Train Operation |
| ATP | Automatic Train Protection |
| AWS | Automatic Warning System |
| BBN | Bayesian Belief Network |
| BS | British Standards |
| BT | Bombardier Transportation |
| BTLUP | Bombardier Transportation London Underground Projects |
| CCS | Control Command and Signalling |
| CCTV | Closed Circuit Television |
| CENELEC | European Committee for Electrotechnical Standardisation (English translation) |
| CIS | Customer Information Systems |
| CM | Coded Manual (mode of operation of the new train) |
| CRMS | Cable Route Management System |
| CSA | Change Safety Analysis |
| CSDE | Correct Side Door Enable |
| CSP | Current Safety Performance |
| CSPSSL | Current Safety Performance SubSurface Lines |
| CSPVL | Current Safety Performance Victoria Line |
| DC | Direct Current |
| DRACAS | Defect Reporting, Analysis and Corrective Action System |
| DTG-R | Distance To Go – Radio (signalling system) |
| ECB | Engineering Change Board |
| ECR | Engineering Change Request |
| EDF | Électricité de France (Energy Provider company) |
| EEPL | EDF Energy Powerlink Limited |
| ELLCCR | Extra Low Loss Conductor Rail |
| EMC | Electro-Magnetic Compatibility |
| EMI | Electro-Magnetic Interference |
| EN | European Norm |
| ERTMS | European Railway Train Management System |
| ESAC | Engineering Safety and Assurance Case |
| ESAP | Engineering Safety and Assurance Process |
| ESM | Engineering safety Management |
| ETCS | European Train Control System |
| EU | European Union |
| FET | Fault-Event Tree |

| Acronym | Definition |
|---|---|
| FMEA | Failure Modes and Effects Analysis |
| FMECA | Failure Modes, Effects and Criticality Analysis |
| FRACAS | Failure Recording, Analysis and Corrective Action System |
| FSP | Final Safety Performance |
| FSPVL | Final Safety Performance Victoria Line |
| FT | Fault Tree |
| FTA | Fault Tree Analysis |
| FV | Fussel-Vesely (importance value) |
| GLEE | General Loss Estimation Engine |
| GPAD | General Parametric Data Set |
| GSN | Goal Structuring Notation |
| GUI | Graphical User Interface |
| HAZID | HaZard IDentificaiton |
| HAZOP | HAZard and OPerability (study) |
| HF | Human Factors |
| HMI | Human Machine Interfaces |
| HSE | Health and Safety Executive |
| ICSA | Initial Change Safety Analysis |
| IEEE | Institution of Electrical and Electronic Engineers |
| IET | Institution of Engineering and Technology |
| IHRG | Interdisciplinary Hazard Review Group |
| INCA | Incident Capture and Analysis database |
| INCOSE | International Council on Systems Engineering |
| ISA | Independent Safety Assessor |
| ISAE | Integrated Safety Assurance Environment |
| ISP | Interim Safety Performance |
| ISPVL | Interim Safety Performance Victoria Line |
| IT | Information Technology |
| LUL | London Underground Limited |
| LVAC | Low Voltage AC |
| MA | Manned Automatic (mode of operation of the new train) |
| MoP | Member of Public |
| MR | Metronet Rail |
| MRBCV | Metronet Rail Bakerloo, Central and Victoria Line |
| MSCIP | Manchester South Capacity Improvement Project |
| NDUP | Neasden Depot Upgrade Project |
| OIDB | Objects and Interfaces DataBase |
| OPO TT | One Person Operation Track to Train Closed Circuit |

| Acronym | Definition |
|---|---|
| CCTV | Television system |
| PAD | Parametric Data Set |
| PD | Position Detector |
| PDD | Project Definition Document |
| PFI | Private Finance Initiative |
| PHA | Preliminary Hazard Identification |
| PKP | Polskie Koleje Panstwowe (Polish railway authorities) |
| PM | Protected Manual (mode of operation of the new train) |
| PMF | Project Management Framework |
| PPP | Public Private Partnership |
| PSR | Permanent Speed Restriction |
| PTI | Passenger Train Interface |
| QRA | Quantified Risk Assessment |
| RAM | Reliability, Availability and Maintainability |
| RAMS | Reliability, Availability, Maintainability and Safety |
| RBD | Reliability Block Diagram |
| RM | Route Manual (mode of operation of the new train) |
| RSF | Right Side Failure |
| RSSB | Railway Safety and Standards Board |
| SAA | Station Area Accident |
| SCC | Service Control Centre |
| SCID | System Context and Interface Diagrams |
| SDO | Selective Door Opening |
| SER | Signalling Equipment Room |
| SHL | System Hazard Log |
| SHWW | Sandbach/Wilmslow (geographical area of railway) |
| SIL | Safety Integrity Level |
| SM | Slow Manual (mode of operation of the new train) |
| SSC | System Safety Case |
| SSL | Sub-Surface Lines |
| SSR | Sub Surface Railway |
| SUP | Subsurface Lines Upgrade Programme |
| TEN | Train European Network |
| THR | Tolerable Hazard Rate |
| TPWS | Train Protection and Warning System |
| TSI | Technical Specification for Interoperability |
| TSR | Temporary Speed Restriction |
| UNISIG | (European) Union Industry OF Signalling |

| Acronym | Definition |
|---|---|
| VAF | Value of Avoiding a Fatality |
| VL | Victoria Line |
| VLU | Victoria Line Upgrade |
| VLUP | Victoria Line Upgrade Programme |
| VPF | Value Preventing Fatality |
| WCMU | West Coast Management Unit |
| WRI | Wheel Rail Interface |
| WRSL | Westinghouse Rail Systems Limited |
| WSF | Wrong Side Failure |

**Table 1-1: Acronyms and Abbreviations**

# CHAPTER TWO

# PROBLEM DEFINITION

## 2.1 Chapter Introduction

This chapter presents the findings of the research against the history of the perception and understanding of uncertainty and risk, as well as the investigation of the existing theoretical and analytical framework in relation to the treatment of safety in engineering.

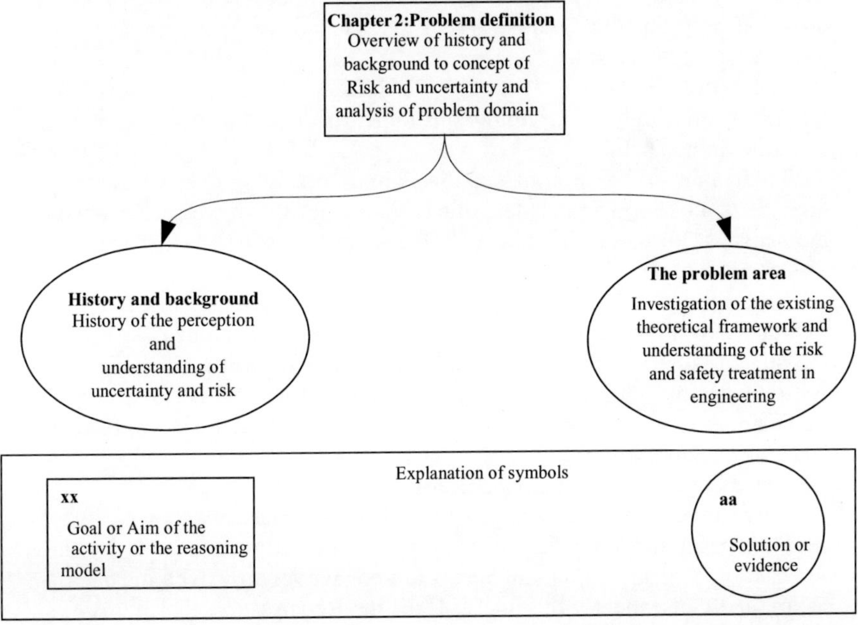


Figure 2-1: Structure of Chapter 2

## 2.2 History and Background (Waldrop, 1992), (Bernstein, 1996)

Everyday experiences are rich with uncertainties and (most of the time unconscious) risk analysis. The treatment of risks that we are accepting every day familiarises us with the subject. Impulses of nature, inaccuracy of our senses and tools, new technologies and the foolishness of human beings all complement the level of uncertainty. Early in life, we learn to rely on a series of intuitive models of common situations, the outcomes of which depend on unknown factors.

This ability of our mind to perform sometimes complex statistical analysis is often sufficient. Still, situations where one could benefit from a more sophisticated treatment of issues are many.

Unfortunately, ignorance about the scientific advances in this field, as well as far too much confidence in intuition, prevents one from gaining benefits by using powerful and, more often than not, simple sets of tools provided by probability, statistics and the sciences of dynamical systems and stochastic processes.

The word "RISK" evolves from *"RISKARE",* the Latin for "TO DARE". Amusingly, if we follow this logic, risk is a choice we make, not a predetermined path.

To explain the creation of universe, Greek mythology used a game of dice. Zeus, Poseidon and Hades rolled the dice for the universe. Zeus won the heavens, Poseidon the seas and Hades ended up with hell, becoming master of underworld.

Regardless of the fact that risk taking has been implanted in our existence from the beginning, development of the science of risk, and of statistics, has been somehow delayed when compared with other sciences. Astronomy, medicine, philosophy, physics and mathematics all have foundations in great ancient cultures of Egyptian, Persian, Greek, Roman and Chinese civilisations. On the other hand, the first serious study of risk happened during the Renaissance (Bernstein, 1996).

There are two main reasons for this long delay (Sterman, 2000).

Firstly, for too long the belief was that the future is shaped by the forces of gods, and that human beings are not actively involved in shaping nature and consequently the future. Until the Renaissance, the future was regarded as an already written book. Fate was determined by the sins of the past and there was nothing human beings could do to change it. People's perception of the future was passive.

Depending on different religions and cultures, for people of ancient times, the future was either a matter of luck or the result of the closeness of their sins to their god's wish. For thousands of years people's lives were very static, and the most important variable influencing their lives was the weather and other unforeseen natural elements.

The second reason is the fact that the purpose of the numbering system at the time was to satisfy only the needs for recording of measurements, but not for the calculations. Then Hindus, in about 500ΑD, developed the numbering system we use today. Arabs adopted this numbering system and spread it across the western civilisation through Spain. The centrepiece of the Hindu-Arabic system was the invention of zero. This supported rapid development of calculus and algebra.

Modern mathematics is based on the work of Aristotle and the philosophers who preceded him. Their work has culminated in several "Laws of Thought", sets of theories supporting concise and systematic approach to logic and later mathematics. One of the most important laws was the "Law of Excluded Middle". The law stated that every proposition must be either True or False. Every statement is True or False, 1 or 0.

Nevertheless, the ancient Greeks gave a thought to probability theory. Socrates defines the meaning of the Greek word "Eikos" (Εικοσ̤as "likeness to truth", and reveals an important phenomenon, "likeness to truth is not the same thing as the truth". Plato, in his work, indicated that there was a third region, between True and False, a space where nothing was completely True or False, but that was as far as the ancient Greeks went.

Only when the civilisation reached a phase of progress that required humans to take charge of their own existence in order to secure further advancement and control over the ever increasing pace of change, did the need for the prediction of future events became a necessity. During the renaissance, the pace of social, cultural and environmental change increased dramatically. People could not afford to be at the mercy of fate any more, they could no longer remain passive in the face of an unknown future.

The intensifying pace of change and competition all around, forced the transformation of mysticism into science and logic, and the human race made the first attempt to liberate itself from self-imposed limitations.

The mathematical foundation of the concept of risk, the theory of probability, is a result of joint efforts of two great minds of the time (1654), Blaise Pascal, a French mathematician and Pierre de Fermat a lawyer and mathematician. At about the same time, Antoine Arnauld, a theologian, was first to state that "Fear of harm ought to be proportional not merely to the gravity of the harm, but also to the probability of the event", or in other words, only the combination of likelihood of outcome and value of outcome should influence a decision.

Methods of statistical sampling and the Law of Large Numbers were invented in 1703 by Jacob Bernoulli. Abraham de Moivre discovered the concept and structure of normal distribution in 1730. Around the same time Daniel Bernoulli defined the systematic process by which people make choices and decisions. Thomas Bayes, in about 1750, discovered a way to make better-informed decisions by mathematically blending new into old information. In 1875, Francis Galton discovered regression to the mean, and Harry Markowitz demonstrated mathematically why diversification is desirable in 1952.

Everyone knows that "uncertainty" affects a project manager who is about to define the price for the contract, an engineer about to sign replacement approval for a piece of equipment, or a trespasser taking that first dangerous step. The concept of hazard comes about due to our recognition of future uncertainty, our inability to know what the future will bring in response to a given action today.

Uncertainty implies that a given action, or non-action, has more than one possible outcome.

'Uncertainty' in everyday use denotes unknown factors that influence the outcomes of observed developments. To analyse these more efficiently, one needs to consider several fundamentally different types of uncertainty (Waldrop, 1992), (Checkland, 1984), (Bhatnagar and Kanal, 1991).

The first category can be described in an *Objective Probability Framework*. This is suitable, for example in a Quality Control Methods context, where we monitor samples and ensure that required standards are met. We also frequently study the likely impact of changes in systems to see how we might improve efficiency or reduce costs.

In the second category we will include all the uncertainties that arise at a fundamental level, that is, the uncertainties that arise because parameters of the problem are non-computable or are fundamentally non-deterministic. These kinds of problems are found most notably in the theory of computation and in quantum mechanics[1] respectively.

Some problems of uncertainty may not be easily classified into either of the two categories, possessing instead the elements from each. We are

---

[1] These two types of problems can be argued to be formulations of the latter (quantum mechanical) on different levels of complexity, i.e. the fact that non-computable problems exist may well be a direct consequence of the existence of non-deterministic ones. This is possible since quantum mechanics is intended to be description of nature at the most fundamental level.

able to predict, with significant accuracy, safe operating limits for railway system. Here we deal with uncertainties of the first kind. Still, acts of terrorism, human error or foolishness, and the unpredictable nature of some mechanical failures all introduce uncertainties of the second category into our analysis. There is a third type of uncertainty that is introduced as a consequence of our ignorance of background theory, either conscientiously because it is impractical or unjustifiably costly to include these considerations into calculations, or as a result of an unawareness of all parameters or laws governing the outcome. In these situations, one chooses to treat the problem as one of the first two types.

The third kind of uncertainty deals with potential hazards and exploring the unknown. It is often a subjective, rather than an objective, exercise, and our perceptions are an integral part of our assessment. The second type, we will use to classify all the problems where uncertainty occurs as a result of an essential difficulty in calculating the results of a deterministic process. This includes chaotic systems (ones in which the outcome is deterministically dependent on initial conditions, but is also extremely sensitive to their change, making it practically impossible to predict it) and problems in which computing a solution is probably unfeasible.

Problems of this class can be often met in a number of engineering fields, including certain behaviours of a realistic pendulum, the rate at which a fluid is dripping from a tap or a problem of calculating the most efficient ways of traversing complex networks as a model of routes taken by electric shocks, and so on.

In technical applications, uncertainty can be categorised in three classes (Fenton and Hill, 1993):

1. Incompleteness of relevant knowledge;
2. Uncertainty of knowledge;
3. Imprecision of knowledge.

In most situations, the relevant knowledge is incomplete. In these cases one is faced with the problem of deciding which one of potentially many different solutions is the true one.

More often than not we are not sure that the information we have is the absolute truth.

Uncertainty of data relates to the observations of nature or society. The observer is uncertain about what has actually been observed, or about the measurements taken.

Measurement and sampling are two sources of data uncertainty. There is always some uncertainty in any measurement, because of the limited precision of the measuring device. This can usually be determined from the characteristics of the device, and by the repeated sampling and statistical characterisation.

Imprecision of knowledge or rule uncertainty relates to reasoning about the observations. The observer is doubtful about the conclusions drawn from the data.

When considering systems, as defined in the section "Definitions" (1.4), the uncertainty is related to incompleteness, uncertainty and imprecision of knowledge about systems constituents, or objects that make up a system, the connections between these objects, the organisation of objects, the system topology, the system environment and interactions between the system and its environment.

Most commonly, practical problems have uncertain factors of more than one type. In real life the number of parameters influencing outcomes of all but the most trivial of problems are so numerous that we usually decide to ignore most of them and describe them as 'uncertainty' in the problem, regardless of the type to which they belong.

## 2.3 The Problem Area – detailed analysis

"Fear of harm ought to be proportional not merely to the gravity of the harm, but also to the probability of the event" (La logique, ou l'art de penser (Logic, or the art of thinking); 1662 – Antoine Arnauld as cited in (Bernstein, 1996)). Therefore 'safety' can be defined as freedom from harm.

Before it is possible to proceed further, it is important to describe our world from the point of view of the person dealing with the uncertainty:

1. Our world is speculative. "Our knowledge of the way things work, in society or in nature, comes trailing clouds of vagueness. Vast ills have followed a belief in certainty". (Kenneth Arrow as cited in (Bernstein, 1996));
2. Perfection implies repetition of assessment, estimation, calculation, evaluation, analysis and computation (Fenton and Hill, 1993);
3. Precision is a measure of vicinity of the measurement to the correct value (Fenton and Hill, 1993);
4. Both, the gauging of the tools used for measurements as well as the dimensions measured fluctuate constantly (Fenton and Hill, 1993);
5. A degree of required precision, or in other words 'allowed level of uncertainty', depends on the criticality of the measure to be used (Hessami, 1999b);

6. Risk is directly proportional to uncertainty, the lesser the uncertainty, the lower the risk. Furthermore, the more we know about the world the lesser is the uncertainty;
7. Risk and time are interrelated. Without time there is no risk. The risk is only relevant in relation to future. Passage of time transforms the risk and its nature.

Therefore, in the simplified sense, every action is "uncertain", from crossing the street to building a railway system. The term "hazard" is usually reserved, however, for situations where the range of possible outcomes to a given action is in some way significant. A significant outcome is a very loose definition, but in truth it is a highly flexible concept. The level of significance varies depending on the level of gain and the human perception of acceptable uncertainty.

Common actions like crossing the street are usually not considered "hazardous" while building a railway system can involve significant uncertainty. Somewhere in between, actions pass from being non-hazardous to hazardous. This distinction, although vague, is important; if one judges that a situation is hazardous, the level of hazard becomes one criterion for deciding what course of action should be pursued.

Existing methodologies for assessment and management of the safety risk on railways are mostly empirical and have been developed out of a need to satisfy the regulatory requirements and in response to a number of major accidents. Almost all of these processes and methodologies have been developed in support of approvals of specific products or very simple systems, and do not add up to a holistic, coherent methodology, suited for the analysis of modern, complex systems, involving many vastly different constituents (software, hardware, people, products developed in different parts of the world, etc.). Thus, the complexities of modern railway projects necessitate a new approach to risk analysis and management.

At the outset, the focus is on the organisation of the family of existing system analysis methodologies into a coherent, heterogeneous methodology. Relationships between different methodologies and their properties were investigated with the aim of defining the rules for the embedding of models into a hierarchical nest, and relating their emergent properties.

Once an outline framework had been created through utilising the existing methodologies on real life projects, the research focused on the development of methodologies in support of the activities not catered for, or not sufficiently supported by, existing methodologies, and integrating these new methodologies, with existing ones, into a new coherent process.

In general any activity is undertaken (Hessami, 1999b) when there is a perception that the balance between the potential loss and the potential gain is such that it is worth pursuing the activity further. At that point, some form of risk analysis becomes viable. Most authors define the risk as a combination of loss and frequency or probability of occurrence. However, (Hessami, 1999b) other parameters define the risk as well:

1. Nature of the consequent loss;
2. Severity of the loss;
3. The nature of initiating hazard;
4. Affected party;
5. Probability or frequency of occurrence of the consequence.

The detailed analyses of these can be very profitable. This, 'more detailed analysis' can extend to several levels to include the above-mentioned parameters, but the most common, as well as the simplest, is the parameter usually covered by probability and statistics.

In engineering problems, detailed analyses of risk and its attributes can lead to significant savings in time and money (e.g. analyses of essentially chaotic behaviour of winds lead to notable improvements in designs of buildings and bridges, not to mention benefits of weather forecasting, a classic example of a chaotic system).

Three high level principles of the analysis of safety risks have been identified that must be respected and supported by adequate processes, because only by adhering to these principles will one be able to:

1. Ensure completeness of analysis;
2. Build a defendable argument in support of the final results (not forgetting that the choice based on the analysis may directly influence decisions potentially affecting human lives and costing millions).

The first principle is that of a Systematic Approach to problem solving. Planning the safety activities, documenting the work and enforcing rigorous configuration control of all important documents and data is paramount. For different problems, different tools and techniques may be appropriate to ensure the completeness of analysis.

A Holistic approach to the analysis is the second principle. From a systems perspective, safety is an emerging property of a system. When analysing the system, all elements of said system must be taken into consideration (software, hardware, human factors, socio-political influences and environmental issues, to mention some) throughout the life cycle of the system. The information about the analysed system should be reliable and complete.

The starting point of the safety management process is usually taken to be the risk assessment. More often than not, when doing risk assessment and analysis, the knowledge of experts in the particular field is exploited if not essential. Nowadays even apparently simple systems consist of parts designed by experts of different disciplines: electronics, mechanical, software and ergonomics come to mind first. On top of this, modern products are used in a wide variety of applications with different operational conditions.

Finally, the third principle of safety risk assessment is the necessity of extensive use of Domain Specific Expertise. The systematic risk based approach to any undertaking has been developed within Railtrack PLC (now Network Rail) and has become a nationwide Engineering Safety Management guidance adopted by LUL and the whole of UK railways. The following are the key risk assessment principles (RSSB, 2007):

1. Hazard Identification;
2. Causal Analysis;
3. Consequence analysis;
4. Loss analysis;
5. Options analysis;
6. Impact analysis;
7. Demonstration of compliance with the bench mark.

The following diagram depicts the above-mentioned process in more detail.

The 7 Stage Process assumes that the system is already well understood and defined.

Several approaches to hazard identification exist (for example, guidance in (Civil Aviation Authority, 2006)), but the most common is in a form of a structured brainstorm where the focus of the session is secured by consideration of the keywords, such as NOT, LATE, EARLY, ALSO, etc, on the problem domain.

The result of *hazard identification* is a linear mass of information. Hazards are defined on their own without any reference to potential relationships and dependencies to other hazards. It would be extremely difficult to analyse each of these hazards on its own and to simultaneously ensure that all interactions between different hazards and common dependencies of multiple hazards have been systematically analysed.

The final outcome of the analysis must not include any double counting, or exclude any of the constituents of the system.

This is why the grouping of hazards of the same origin into larger entities referred to as core hazards, should be performed as the next step of analysis.

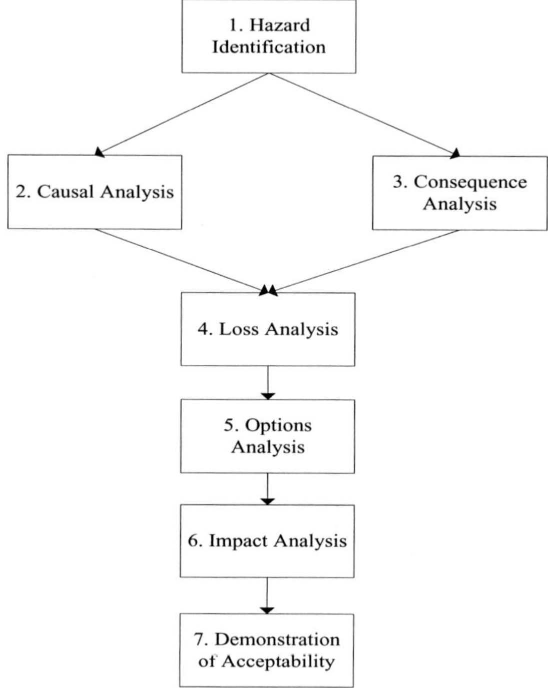


Figure 2-2: 7 Stage Process

On-line capture, using the projector and the laptop, of information during hazard identification sessions (and any other group workshops) has been proven to be extremely effective. The main benefit is that the panel can agree (or argue about, which is more usual) the final version of information to be captured (e.g. the wording of the hazard and its consequences and mitigations, in the case of hazard identification). Hazard identification sessions must be well structured and planned in advance. All participants must be allowed to express their opinion and encouraged to take an active part in the session.

Domain Competence Requirements are determined by the problem under consideration. Care must be taken to ensure that expertise, related to all aspects of the problem area, is comprehensively incorporated into a study throughout the process. In order to secure as comprehensive as possible a source of data and ideas, care should be taken to form groups with wide range of expertise that fits the problem space.

The aim of *causal analysis* is to identify all the events/causes, and all the intermediate conditions, which lead to a hazardous condition.

*Consequence analysis* involves establishing the intermediate conditions and final consequences that may arise from a hazard. It involves a bottom-up assessment of each hazard and is focused on the post hazard horizon. At each intermediate state, existing defences against potential escalation should be identified. The defence may be equipment, procedure or circumstance. The sequence of the identified intermediate conditions is termed "the hazard development scenario". Final consequences fit into one of the following categories: predominantly safety related consequence, predominantly commercial consequence, predominantly environmental consequence, broadly safe condition.

*Loss estimation* is the process of estimating the losses that result from an accident or incident. These losses may be of three kinds: commercial, safety and environmental. Under each of these categories there are a number of ways in which losses may be incurred. These are termed loss mechanisms.

Each of the loss mechanisms should be considered. Therefore, the first task is to identify each of the relevant loss mechanisms for a particular event.

Having identified the relevant loss mechanisms, it is then necessary to estimate the associated losses. It is useful to rank the loss mechanisms in order of estimated magnitude, so that it is apparent if the losses are primarily commercial, safety or environmental.

Safety losses in terms of injuries or equivalent fatalities are notoriously difficult to estimate.

For major accidents, there is (fortunately) not a large enough sample for results to be statistically significant, so often it is necessary to make a (conservative) estimate.

In order to normalise the units of loss (commercial, safety and environmental) and enable comparison of risks and gains, the safety loss is often converted into financial losses. In UK the Health and Safety Executive's (HSE) Value for Preventing a Fatality (VPF) is often used. This is £0.99 million for one or less equivalent fatalities and £3.12 million for accidents involving more than one equivalent fatality, where 1 equivalent fatality = 1 fatality or 10 major injuries or 200 minor injuries. Safety losses are, by convention expressed in terms of Potential Equivalent Fatalities per annum.

*Options analysis* is aimed at identification of potential reliability improvements and damage containment against various system or procedural configurations. This stage of the process is a creative phase, employing empirical knowledge and inspiring elicitation techniques. For each identified option, the following information should be elicited:

1. Costs;
2. Loss Prevention Capability;
3. Resource & Competency Requirements;
4. Time-Scales;
5. Uncertainty & Dependency;
6. Impact on Other Loss Dimensions.

The systematic assessment of the overall effect of each identified option on the potential risks is referred to as the 'impact analysis'.

The *Impact* of each identified option on the reduction or containment of identified risks should be assessed.

In case of safety risk, the demonstration of compliance in the UK involves comparison of the total individual risk arising from the malfunction of the system against the industry benchmark as a minimum required safety performance and against the investment required to improve the system further.

The first part of this stage is related to the maximum acceptable risk exposure which in UK is set by HSE (Health and Safety Executive, 2009).

The second part of this stage is related to the foundation of the assurance that everything reasonable has been done in order to minimise the risk arising from the malfunction of the system. Each of the identified options should be assessed, the improvement that the option could bring if it is implemented should be weighed against the cost of the implementation, as well as other relevant parameters.

The risk assessment can be qualitative or quantitative. Once all the reasonably practicable options have been implemented, it can be demonstrated that the safety risk arising from the malfunction of the system is acceptable.

Current practice in the UK demands demonstration of compliance with the 'As Low As Reasonably Practicable' (ALARP) principle for any undertaking. This demonstration is typically structured in the form of a Safety Case.

Very similar approaches to safety management are discussed in many other papers and standards ((SAMRAIL, 2004) to mention one), but all of these are very similar to the "7 stage process" and to avoid duplication these will not be discussed here.

The extension of the "7 stage process" is the rational approach to risk assessment, inclusive of both the rewards and risks of the enterprise, as well as the benefits and the detriments brought by the system. Following this logic (Hessami, 1999a), if a hazard is the precursor to risk, the opportunity is the precursor to reward. Therefore, in order to ensure that the totality of the system has been analysed (for safety) it is necessary to assess the benefits of the system as well as risks.

The safety assurance of a system covering the complete generic lifecycle could be split into two phases (CENELEC 1998), (CENELEC 2003), (Hessami, 1999b), (RSSB, 2007), see Table 2-1:

1. Safety Engineering, covering design, development and realisation of the system;
2. Safety Management, covering deployment, maintenance, retrofit and decommissioning.

| Generic lifecycle phase | Principal ESM activities |
|---|---|
| Concept and feasibility | Preliminary hazard identification |
| | Establish the hazard log |
| | Define the preliminary safety plan |
| Requirements definition | Hazard Analysis (and revisiting of hazard identification) |
| | Risk Assessment |
| | Establish safety requirements |
| | Define full safety plan |
| Design | Risk Assessment |
| | Safety Audit |
| Implementation | Risk Assessment |
| | Safety case |
| Installation and handover | Safety Assessment |
| | Safety Endorsement |
| | Transfer of safety responsibilities to the user |
| Operations and Maintenance | Update the Hazard Log and safety case |
| Decommissioning and Disposal | Update the Hazard Log and safety case |

**Table 2-1: Principal engineering safety management activities.**

The safety engineering phases correlates with the pre operational phases whilst the safety management phase correlates with the application and maintenance aspects of the system lifecycle. Although the Safety

Engineering and Safety Management phases are interconnected and overlap in many aspects, the methodology for management of risks, and the tools and processes supporting the risk management are quite different.

During the safety engineering phase, the focus of safety risk management is on the provision of a system with the potential to provide a safe performance. The focus of the safety management phase is to realise that potential. As part of the railway's Engineering Safety Management (ESM) guide (RSSB, 2007), a detailed breakdown of activities is provided and related to the system life cycle that is required to ensure a safe delivery and use of a system.

Most of the existing processes and guidelines state what needs to be a result of the safety risk analysis, broadly outlining the expectations from each of the above listed and depicted activities in Table 2-1. Alternative methodologies exist for some of the activities, but not for all, each one invented and used in the context of different aspects of systems structure, behaviour, and/or its emerging properties.

However, none of these amounts to an integrated framework for the systems safety risk analysis and management. Many examples of a failure to assure the clients and relevant authorities (Her Majesty's Railway Inspectorate, for example) of the satisfactory safety performance of the system exist in the railway industry. To mention one example, the failure to demonstrate the satisfactory safety performance of a novel signalling system (the upgrade of a railway junction on a main railway line) led to two year long delays in implementation and a mounting cost (some estimates are that the final cost of the system will be two orders of magnitude higher then original tendered cost).

There are also examples of the opposite, where a grossly underestimated cost of an option may have caused a delay to the implementation of what is potentially an almost equally efficient option. Following on from the investigation into the Clapham junction train accident in 1988 (35 fatalities and 69 major injuries) lead by Sir Antony Hidden, a recommendation was made that the Automatic Train Protection system should be fully implemented within 5 years of selecting a specific type of the system, with high priority being given to densely trafficked lines (Department of Transport, 1989), (Health & Safety Commission, 1998). Subsequently, the recommendations from Hidden enquiry were reinforced in particular in the report into an accident at Cowden in 1994. However, the manner in which technology was developed, and consequently, the high cost of the development and implementation at the time when the recommendations were made, prevented the implementation of the recommendations within the timescales suggested by the various enquiries. Furthermore, the spiralling cost of the failed implementation of the modern ATC (ERTMS/TCS) system along the West Coast Main Line was a major cause of the demise of Railtrack. During the late 90s (Health & Safety Commission, 1998) TPWS was identified as a reasonable, practicable minimum requirement. The rollout of this system has been successfully initiated in parallel with the conclusion that introduction of the ATP system may not be justified on the bases of cost until the technology develops further.

If at the time of the first recommendation a more measured view was taken of the whole life cost of ATP systems, TPWS could have been proposed as a pragmatic way forward and implemented across the most critical parts of the network much earlier, thus potentially preventing some of the later accidents. This is an example of where the cost of safety must be analysed systematically and carefully, taking into account the full life cycle cost, to support the justifiable decision-making.

Furthermore, in different industries, significantly different levels of investment are expected on health and safety. Thomas and Stupples (Thomas and Stupples, 2006) have devised a new approach to the appraisal of reasonable practicability as part of an options and impact analysis. The J-value is the ratio of the actual spend against the maximum reasonable spend, which in turn is a function of life expectancy, average income and work life balance. A number of case studies have been undertaken, including an assessment of J-value related to implementation of TPWS and ERTMS ATP systems. For both (TPWS and ERTMS ATP) the J-value was found to be greater than unity, indicating that both are costing more than it should. The J-value for TPWS was found to be between 3.8 and 11.3, and for ERTMS ATP between 38 and 132, strongly supporting the argument that ATP systems are far too expensive for the realised benefit.

All of this must be taken into account when options and impact analyses are undertaken.

The root cause of the failure to undertake the complete, holistic analysis is twofold:

1. Lack of understanding of the complexities involved in the development of modern systems;
2. None-existence of a holistic methodology, and the process supporting it, throughout the lifecycle safety analysis and management.

As discussed later in this book, the complexity of modern projects and products demands the Systems approach to the analysis of Safety as an emerging property of the system, for the simple reason that with the increased complexity of the systems produced by human kind, our ability to comprehend the totality of the system without a structured methodical process is decreasing.

## 2.4 Chapter Conclusion

In this chapter, the author outlined the background and the history of the human perception of risk. Since risk is an abstract, almost philosophical concept, this initial research and discussion was necessary to set out the scene for further research. Following on from that, the author investigated the problem domain further, aiming to plot out a direction for the research. The author identified the gaps in existing knowledge and application, and outlined the high level requirements that a future framework must satisfy.

# Chapter Three

# Basic Systems Concept, Modelling Methodology and Background Literature

## 3.1 Chapter Introduction

The findings of the initial literature research are presented here. The research was conducted in two specific areas (as indicated by Figure 3-1 below); research of systems theory relevant to safety engineering and management, and research of the tools and methods used for specific engineering safety problems.

## 3.2 Background to the notion of the system

There are many different definitions of a system.

Benjamin and Wolter (Blanchard and Fabrycky, 1998) conclude that a system is assemblage or combination of elements or parts, forming a complex or unitary whole, such as a river system or a transportation system; any assemblage or set of correlated members, such as a system of currency; an ordered and comprehensive assemblage of facts, principles or doctrines in a particular field of knowledge or thought, such as a system of philosophy; a coordinated body of methods or a complex scheme or plan of procedure, such as a system of organisation and management; any regular or special method of plan of procedure, such as a system of marking, numbering or measuring. A system is characterised by unity, a functional relationship, and a useful purpose.

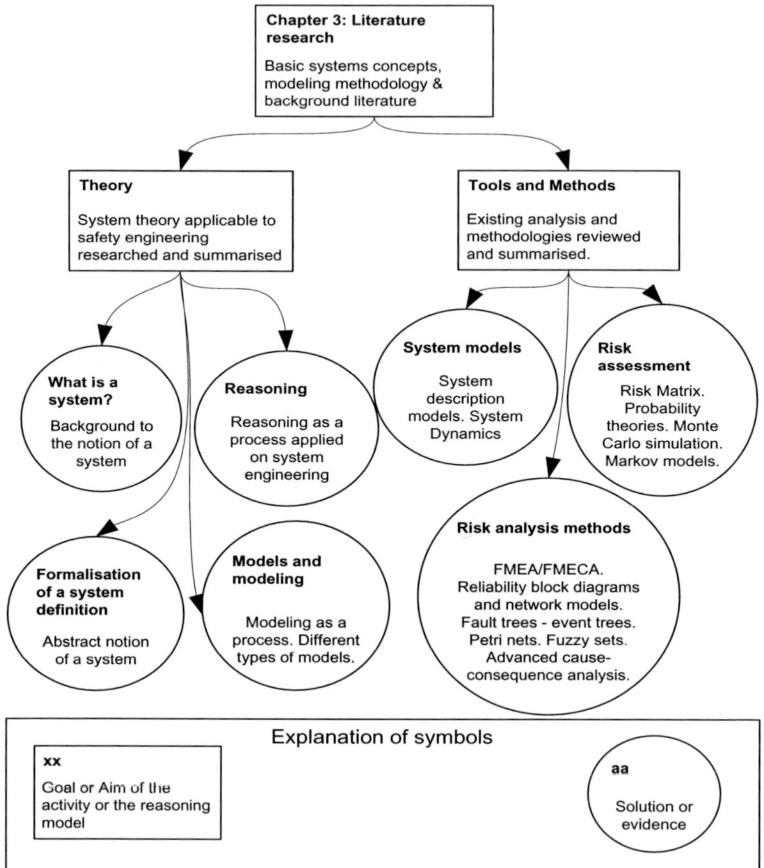


Figure 3-1: Structure of chapter 3

A system is a set of interrelated constituents working together toward some common objective or purpose. The set of constituents has the following properties:

1. The properties and behaviour of each constituent of the set have an effect on the properties and behaviour of the set as a whole;
2. The properties and behaviour of each constituent of the set depend on the properties and behaviour of at least one other constituent in the set.

3. Each possible subset of constituents has the two properties listed previously; the constituents cannot be divided into independent subsets.

A system is more than a sum of its parts. However, the constituents of a system may themselves be systems, and every system may be part of a larger system in a hierarchy.

Another, more pragmatic definition of a system, comes from the internal document of an engineering company (Blanchard and Fabrycky, 1998): “A system is the interacting combination of equipment, people and processes designed to accomplish a defined objective. The environment is that part of the rest of the world that a system interacts with.”

As a final example, the following definition of a system is given by Norman Fenton and Gillian Hill (Fenton and Hill, 1993): “A system is an assembly of constituents connected together in an organised way and separated from its environment by a boundary. This organised assembly has an observable purpose, which is characterised in terms of how it transforms inputs from the environment into outputs to the environment. In order for systems to function effectively they generally require some kind of control mechanism. This monitors external behaviour, providing feedback and enforcing changes to the system where necessary.”

In his book (Waring, 1996), Alan Waring defines the system in a similar manner to the above-mentioned authors.

Any system is characterised by its emerging properties, and the properties that are an outcome of the joined act of the system parts. The system constituents (Blanchard and Fabrycky, 1998) are the operating parts of a system consisting of input, process, and output. Each system constituent may assume a variety of values to describe a system state, as set by some control action and one or more restrictions.

Attributes are the properties of discernible manifestations of the constituents of a system. These attributes characterise the system. Relationships are the links between constituents and attributes. A purposeful action performed by a system is its function.

A system is defined by its boundary or scope. Everything that is outside the boundary is considered to be the environment. No system is completely isolated from its environment. Material, energy, and/or information are passing through the boundaries as input to the system. In reverse, the material, energy, and/or information that is passing through the boundaries from the system is referred to as output. That which enters the system as one form, and exits from system in another form, is known as throughput.

Every system is made up of constituents, and in turn, any constituent may be broken down into smaller constituents. If two hierarchical levels are involved in a given system, the lower is conventionally called subsystem.

A subsystem (Fenton and Hill, 1993) is a system in its own right that is contained within some other system. Most real systems are so complex that the only way one can understand them is to understand their recursive structure by thinking in terms of their simpler interconnected subsystems.

If these systems are themselves too complex most often one again thinks in terms of their subsystems. This process of looking for simpler subsystems within subsystems is repeated as many times as necessary. The process is called top down decomposition, and it gives rise to notion of levels of abstraction. It is the only known means of analysing complex systems.

Some authors have made an attempt to classify systems. Benjamin and Wolter (Blanchard and Fabrycky, 1998) introduce the following classification:

1. Natural and human made systems;
2. Physical and conceptual systems;
3. Static and dynamic systems;
4. Closed and open systems.

Checkland (Checkland, 1984) supports the idea of a 9 level hierarchy of real world complexity:

| Level | Characteristic | Examples | Relevant disciplines |
|---|---|---|---|
| Structures, frameworks | Static | Crystal structures, bridges. | Description, verbal or pictorial in any discipline. |
| Clock-works | Predetermined motion – equilibrium | Clock, machines, the solar system. | Physics, classical natural science. |
| Control mechanisms | Closed loop control | Thermostats. | Control theory, cybernetics. |
| Open Systems | Structurally | Flames, | Theory of |

| Level | Characteristic | Examples | Relevant disciplines |
|---|---|---|---|
| | self-maintaining | biological cells. | metabolism (information theory) |
| Lower organisms | Organised whole with functional parts, blue printed growth, reproduction | Plants. | Botany. |
| Animals | A brain to guide total behaviour, ability to learn | Birds and beasts. | Zoology. |
| Man | Self-consciousness, knowledge of knowledge, symbolic language. | Human beings. | Biology, psychology. |
| Socio-cultural systems | Roles, communication, transmission of values | Families, clubs, nations. | History, sociology, anthropology, behavioural science. |
| Transcendental systems | Inescapable unknowables | The idea of god. | |

**Table 3-1: Hierarchy of real world complexity**

The author suggests the use of the above classifications to aid the choice of a method for system analysis. One other categorisation of complexity defines two types; essential and accidental.

Essential complexity, as named, is in the essence of the system. It is an inherent part of a system and cannot be eliminated, instead only minimised. Accidental complexity is not the "natural" attribute of the system, but is a consequence of an accident. This type of complexity can be eliminated. The following systems analysis methodologies are suggested (Checkland, 1984), (Waring, 1996), (Blanchard and Fabrycky, 1998):

A. Hard system thinking including following steps:
1. Doing the groundwork;
2. Gaining awareness and understanding of the perceived problem;
3. Establishing overall goals and set of objectives;
4. Finding ways to reach objectives;
5. Devising assessment measures;
6. Modelling;
7. Evaluation;
8. Making a choice;
9. Implementation.

B. Soft system thinking including following steps:
1. Data collection;
2. Analysis;
3. Relevant systems and root definitions;
4. Conceptual modelling;
5. Comparisons to provide debating agenda;
6. Discussing the agenda with the actors;
7. Action for change;

C. Systems failure thinking including following steps:
1. Describing the failure situation;
2. Comparison with paradigms;
3. What do comparisons mean;
4. Learning.

With the aspiration of uniting the different approaches, soft, hard and systems failure systems thinking, the author will introduce a new generic definition of a system.

## 3.3 The abstract notion of the system

A system is a collection of interconnected, organised parts, embedded in the environment that forms a whole, which exhibits some new properties that none of the constituents possess on their own.

From the point of view of a high level safety analysis strategy, if the definition of a system given above is adopted, the distinction between soft/hard systems, and system failures, is neither a natural nor a useful one. Any system, regardless of its attributes and nature, must be analysed in a systematic way and the generic approach should be the same.

Following on from that, one can conclude that the methodology for system analysis should be the same for all systems, although the tools for its analysis may be different. Therefore it would be much more useful to classify systems based on their attributes, thus enabling the selection of appropriate tools for the analysis.

Through work on several projects (Lucic, 2003b), the following conclusions related to systems analysis have been drawn.

The systems we are faced with are, most of the time, not uniform. Modern high reliability systems are often distributed, and consequently, the performance of their constituents and subsystems differ in many aspects (environment, speed, timing), at different places throughout the system.

Nevertheless, the system must be comprehensively defined to facilitate the identification of all the hazards.

In generic terms, any system is fully defined by its:

1. Environment;
2. Constituents (different authors use different names for the constituents, objects, parts, etc);
3. Topology of the internal and the external interfaces and
4. State of the system.

Environment is defined as the surroundings within which the system works, including people, weather, climate conditions, traffic density, etc. The system can be "self contained" i.e. there can be no interactions between the system and its environment, or the system can be "embedded", i.e. the system receives inputs from the environment and is passing outputs into the environment to fulfil its function. The "embedded" systems can also receive the behaviour rules and disturbances from the environment, and can pass the disturbances into the environment.

Constituents, or objects, are the parts which make up the system; its hardware, software and people. Objects are defined by their attributes which are identifiable, and their possible measurable characteristics (Topintzi, 2001).

Objects can be described as the "atomic transformation machines" that transform inputs into outputs in accordance with some transformative function. The transformative function can be "self contained" i.e. the control mechanism of the transformative function can be entirely internal to the object, or it can be "externally controlled", i.e. the measurement of the output can be passed out of the object and the corresponding control can be fed back into the object.

The structure and organisation of connections, or interfaces, between the constituents themselves, along with between the constituents and the environment, is referred to as the topology of the internal and external interfaces. The state of the objects, and the nature of their interconnections (interfaces) and organisation, defines the structure of the system (Topintzi, 2001).

Constituents may change the way they transform inputs into outputs, they can be in different states or the network of interconnections may change its structure, nature or organisation, hence changing the properties and the state of the system. At any moment in time the system may be operating in different states.

It is not unusual that states are grouped into operational modes. For example, some constituents may be operating in the degraded mode of operation, whilst some parts of the system could still operate in the normal mode of operation. In the case of large distributed systems, some constituents of the system may operate in different states of operation as well. For example, some constituents of the system may be in a preparatory, boot-up stage of operation whilst some of the constituents may be already in full operation.

Formalisation of the notion of a system is offered in (Karcanias, 2003), identifying the following definitions in relation to systems' concepts:

Definition 1: A system is an interconnection, organisation of its constituents, objects that is embedded in a given environment.
SYSTEM ⇔ OBJECTS + TOPOLOGY OF RELATIONS + ENVIRONMENT

Definition 2: An object is a general unit (abstract or physical) defined in terms of its attributes and the possible relations between them.

Definition 3: For a given object, we define its environment as the set of objects, signals, events, structures, and conditions, which are considered topologically external to the object, and are linked to the object in terms of relations with its structure, attributes through interfaces with the object.

Definition 4: An attribute for an object is an identifiable and possibly measurable characteristic of the object.

Interfaces of the system, inputs and outputs of the system, or a system constituent, can be generalised in three categories, each of which can be either transferred to other system constituents or be an output of the system:

1. Energy in this case is assumed to be raw energy, motion, heat, electric power, chemical energy;
2. The second category of input or output is defined as information,

for example about a state of the nearby system constituent, measured speed or temperature, confirmation status, etc.;

3. Finally the third category of input or output is an action. Action is defined as an act of the system or constituent or environment, on a system, constituent or environment, resulting in a change of a status of the action recipient.

In a general sense, any system or a system constituent could be presented by the following diagram:

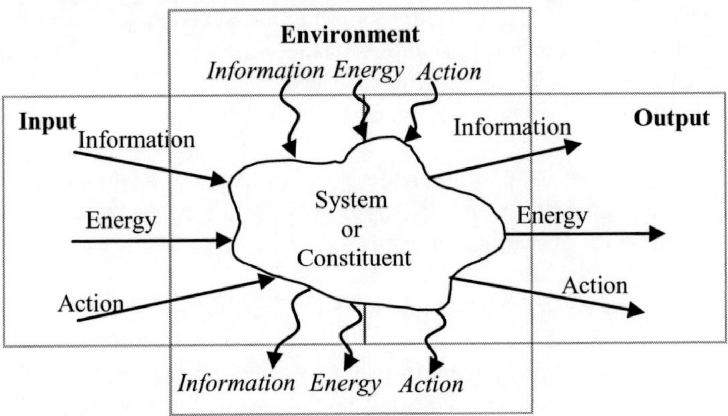


Figure 3-2: Generalised presentation of the system

A combination of the constituents defines the topology of the system, whilst the influence of the environment on the system can be presented as a special case of input/output.

A combination of the constituents' attributes together with the environmental attributes defines all of the emerging properties of the system.

## 3.4 Systems Thinking and Reasoning

An essential part of any problem solving, system analysis and decision-making process, is reasoning.

Reasoning is the process of deducing information about a non-perceptible feature of the problem or non-measurable property of the system, based on the available, measured or observed, information about the problem or parts of the system (Bhatnagar and Kanal, 1991).

In a general form, reasoning activity can be structured as follows:

1 Given/Known:
   1.1 Information/Knowledge about the observable characteristics of a system/ problem;
   1.2 Information/Knowledge about the environment/domain to which the system/ problem belongs.
2 Need to determine:
   2.1 Some information about some unobserved/unobservable characteristics of a system/ problem.

Three concepts are defined in relation to any reasoning:

1. Innate ideas refer to principles or theoretical truths about the nature in which the reasoner believes;
2. The inputs are the experimental observations made by the observer; and
3. The fact net is an interconnection of inputs and innate ideas constructed by means of a given set of relations and operators, it is a network of contingent truths.

The Inquiry System (Portland and Turoff, 2002) is defined as a process followed by someone, either a group of people or something with the aim of creating a network of facts in order to establish the overall truth. The enquirer starts off the Inquiry System with an *assumed* "raw data set". The data set, at the beginning, is considered to describe a characteristic property of the "real world".

This is because, to begin describing the "world" and the enquirer's "knowledge" of "it", the enquirer has to invoke a particular "conceptualisation" or Inquiring System characterisation of "it".

Next, the Inquiry System applies some transformation to the "raw data", creating an input to the model. The model, or the fact net, which is any sort of a structured process, is in fact a set of rules. Rules may be either in the form of an algorithm or a set of heuristic principles.

The model transforms the "input data" into "output information". Finally, the output information may be processed again to put it in the right form so that a decision-maker can use it.

Inquiry Systems are differentiated from one another with respect to the priority given to the various steps of the process or the degree of interdependence assigned to the various steps of the process by each Inquiry Systems.

There are 5 core Inquiry Systems (Linstone and Turoff , 2002):

1. Leibnizian Inquiry System: assumes the existence of an a priori model of the system (the innate ideas or the theory describing the

system), and attempts to arrange the inputs as to suit the a priori model;

2. Kantian Inquiry System: to begin with, this Inquiry System contains a set of independent models built around an independent set of innate ideas and potentially axioms, primitives and rules of inference. The Inquiry System selects a model from the collection and builds a Leibnizian fact net using this model as a base. Using a preset criterion the Inquiry System determines the extent to which the fact net is satisfactory. The model that produces the most satisfactory fact net is the solution to enquiry;
3. Hegelian Inquiry System: is a dialectic Inquiry System. It seeks to acquire the ability to see the same inputs from different points of view. The concept is based on the premise that for each book there is an antibook. The antibook does not have to be a logical negation of the book; it may be simply a book supporting a different view of the same input to the system or different understanding of the effect of the same input on the system. To begin with, this Inquiry System contains a set of independent models built around an independent set of innate ideas and potentially axioms, primitives and rules of inference. The Inquiry System ascertains a model that demonstrates that there is a way to look at reality which fits the selected model. Following that the Inquiry System ascertains the model that supports the antibook. The final step of the Hegelian Inquiry System is development of a "synbook" model of the system, or a model that would examine and resolve the conflicts between the book and antibook models;
4. Lockean Inquiry System (Portland and Turoff, 2002) postulates that the truth content of a system being observed is associated *entirely* with its empirical content. A model of a system is an *empirical model* and the truth of the model is measured in terms of its ability to reduce every complex proposition to its simple empirical referents. The validity of each of the referents is assured by means of the widespread, freely obtained *agreement* between different observers. The accuracy of the model is not dependant on the prior assumption of any theory. The data input is not only prior to the formal model or theory sector but it is separate from it as well. The whole of the Lockean Inquiry System is built up from the data input;
5. Singerian Inquiry Systems: the truth content of a system is relative to the overall goals and objectives of the inquiry. It is said that for these Inquiry Systems the truth is pragmatic. A corresponding fact

net or the model of the observed system is explicitly goal-oriented. The precision of the model is measured with respect to its ability to define systems objectives, to propose alternate means for securing these objectives, and to specify new goals (discovered only as a result of the inquiry) that remain to be accomplished by some future inquiry. It is referred to as a teleological Inquiry System.

Most of the time, the information that is available to the Inquiry System is uncertain, as described later in this report. In these situations, the analyst (Bhatnagar and Kanal, 1991), (Portland and Turoff, 2002) is faced with following options:

1. A Leibnizian approximate Inquiry System is concerned with determining the uncertainty associated with the proposed solution model given the uncertainties related to inputs and the innate ideas, or determining the imprecise version of the solution model such that the solution can be proven to be correct ("true"), given the uncertainties related to the inputs and the innate ideas;
2. A Kantian approximate Inquiry System would attempt to determine the ordered list of several models that provide the best explanation of the inputs. The imprecision associated with the solution model could be used as a criterion upon which to select a preferred solution model;
3. A Hegelian approximate Inquiry System would attempt to determine the pairs of models in support of the book and the antibook. The precision of the information embedded in the models would be then used as a criterion for selection of a solution model.
4. A Lockean Inquiry Systems are the personification of *experimental, consensual* systems. Lockean Inquiry Systems create an empirical, inductive representation of a problem domain, starting from a set of elementary empirical judgments ("raw data," observations, sensations). From these an ever expanding, increasingly more general fact net of the real propositions is created.
5. Singerian inquiry is a nonterminating Inquiry System. Singerian inquirers never give final answers to any question although at any point they seek to give a highly refined and specific response. No single aspect of the system has any fundamental priority over any of the other aspects. The observed system is treated as an inseparable whole. It takes holistic thinking to the extreme and aims to constantly include new variables and additional constituents to broaden the base of concern. It is an explicit postulate of Singerian inquiry that the systems designer is a fundamental part of the

> system, and that, therefore, the system designer must be explicitly considered in the systems representation as one of the system constituents. The designer's psychology and sociology are inseparable from the system's physical representation. Singerian inquirers are the personification of interdisciplinary systems. In a way the Singerian Inquiry Systems are meta- Inquiry Systems, they constitute a theory about all the other Inquiry Systems (Leibnizian, Lockean, Kantian, Hegelian). Singerian Inquiry Systems include all the previous Inquiry System as submodels in their design.

Different models of enquiry are better suited for different problems.

The Leibnizian Inquiry System is an obvious choice in situations where it is known that a good model or theory is available.

If we believe that the truth is equally dependant on the model, and the data, then the most appropriate Inquiry System is the Kantian model.

The Hegelian Inquiry System is most appropriate for ill-structured problems where the truth is more likely to emerge from a clash between a book and an antibook.

The Lockean Inquiry Systems are appropriate for problems where data is available prior to the development of formal theory.

The Singerian Inquiry Systems are best suited for complex system analysis problems where completeness of analysis is paramount.

In addition to above described Inquiry Systems, a number of methodologies for structured group analysis have also been developed. The most common one is known as Delphi.

Delphi is a method for structuring a group communication process in support of an effective technique, thus allowing a group of individuals to deal with a complex problem.

There are two forms of Delphi method. The most common is the paper-and-pencil form, commonly referred to as a "Delphi Exercise." A questionnaire is designed and distributed to a group of experts. The questionnaire is evaluated and the results are distributed to the same group of experts who are given at least one chance to review the original answers. We shall denote this form as *conventional Delphi.* The second form of Delphi technique, sometimes called a "Delphi Conference", that utilises a computer programme to carry out the analysis of the group results. This eliminates the delay caused in analysing the results, turning the process into a real-time communications system.

The Delphi process can be broken into 4 phases. The first phase is the exploration of the subject under discussion. During this phase, each participant freely contributes information pertinent to the subject or system under assessment.

The second phase strives to identify any disagreements between the participants.

The third phase is aimed at resolving the differences and evaluating them. Finally, when all the previously gathered information is analysed, the results are fed back for consideration.

## 3.5 Models and modelling

Modelling is a process of approximating the real world in order to aid representation and understanding of the system, as well as to predict the behaviour of the emerging properties of the system.

Mathematical modelling is a term used to describe the method of approximation of the real world using mathematics (Open University Handbook). A mathematical model is a mathematical relationship linking the variables that represent the possible states of the real system.

All models are characterised by their geometric/logical structure, and if applicable, the algebraic structure that provides the mathematical base of the model.

The two main categories of models are:

1. Descriptive models that support understanding the problem by way of providing structural insight;
2. Decision models that support the decision-making process. The main requirement on these models is to order the preference of one option over another.

Most of the models are causal models. These models are based on the knowledge of causality relationships, which underpin the determination of relating events and the causal structure for a scenario.

Tversky and Kahneman 1982, (Bhatnagar and Kanal, 1991), state "It is a psychological commonplace that people strive to achieve a coherent interpretation of the events that surround them, and that the organisation of the events by schemes of cause-effect relationships serves to achieve this goal."

There are two approaches to modelling:

1. Qualitative models describe the real world, but do not provide any numerical information in relation to model attributes.
2. Quantitative models, on the other hand, approximate the real world using some form of numerical or combined graphical and numerical technique.

The causal structure (Bhatnagar and Kanal, 1991) underlies the observed symptoms of every situation. Most of the time the observable knowledge is incomplete, and consequently, it is difficult if not impossible to uniquely identify the underlying causal model.

In situations like this, one may need to develop a number of different models and then select the most suitable one.

Quantified risk based modelling is increasingly becoming the method of choice for risk assessment. However, risk assessment models are notoriously difficult to validate and verify.

Some scientists believe that it is not possible to verify and validate any model. This argument is based on the fact that natural systems are never closed, and that consequently, models created to represent these systems are never unique. Therefore, according to this school of thought (Saltelli, 2002), models can only be "confirmed" by the demonstration of agreement or non-contradiction between observation and prediction.

Modelling as a process can be split into following steps:

1. The first step of the modelling process is the identification of the problem, who or what is potentially affected by the problem, and what are the parties involved in the problem;
2. Prior to commencing modelling it is desirable to formally describe the system to be modelled. This stage of the modelling process is usually referred to as Conceptual modelling. Conceptual modelling has three distinct phases; elicitation, formalisation (processing) and representation of knowledge. This stage of the modelling process provides a structured approach aimed at gaining an understanding of the problem domain, and the development of the description of the system to be analysed.

Development of the model involves creation of the model to encompass all relevant factors within a logical structure that are capable of simulating the behaviour of the analysed system. This is an iterative process and involves refinement of both the conceptual model and the model itself. In some cases, if the analysed system is complex, it is necessary to produce a hierarchy of models.

There are two types of complexity that may lead to the development of hierarchical models. If the analysed system is complex in terms of its size, number of elements and topological intricacy, it may be prudent to develop a high level model describing the behaviour at the level of subsystems and more detailed models for each of subsystems.

Subsequently, these models can be integrated into a single model or all of the models can be treated as separate entities with the high level model being manipulated using the data derived through detailed model simulations. If the analysed system is complex in terms of the diverse nature of behaviour or the characteristics of system constituents, for which different modelling techniques are required, it may be necessary to use different modelling techniques to model different constituents of the system. Once the analysis is completed, these models can also be integrated.

After the modeller is satisfied with the model it can be used in support of system analysis and the evaluation of different options for a design or a change to be introduced to the system. The important part of this stage of modelling is Sensitivity Analysis, the exploration of uncertainties in the model and how these can be apportioned to different sources of uncertainty in the model input.

## 3.6 Review of existing analysis methodologies

### 3.6.1 System Description Models

Many different methods are used to represent a system. Broadly these may be categorised into two groups:

1. Static models describing the system constituents and their interactions;
2. Dynamic models describing systems behaviour.

Static models mainly describe the system and its constituents. The most frequent model is the System Decomposition Model. System decomposition models often have a hierarchical structure whereby at the highest level only the large subsystems are presented, and then, each of the subsystems is broken down into more detailed presentation and so on. These models are essentially directed graphs, where shapes (nodes) present system constituents, and directed arrows (edges) represent their internal and external interfaces.

Often, these models are used to capture/represent more than just the basic information such as:

1. Type of interface (command/information);
2. Type of system/subsystem constituent (hardware/software/human) and
3. Boundaries between the system and its environment including interactions between the system and the environment.

Examples of these models are presented by Figure 3-3 and Figure 3-4 below.

This type of the system decomposition models is often referred to as the "Bubble Diagram", bubbles represent the subsystems and rectangles represent the constituents at lower level of decomposition. Directed lines are interfaces that link up subsystems and constituents into a system.

A common feature of all decomposition models is that system constituents (constituents or subsystems) are presented by shapes and internal and external interfaces (action, energy or information "carriers") by directed lines, indicated direction of the exchange.

Of the types of dynamic models the State Transition, and Sequence and Collaboration models are seen as the most useful.

The State transition model consists of states (all possible states of the system), triggers (triggers for a system to enter or exit a state) and arrows indicating a direction of a change of a state. An example of a state transition diagram is presented in Figure 3-5 below.

State Transition Models (Pukite and Pukite, 1998) are directed graphs, representing the state space of a system in a given context, the events that cause the transitions and transitions between the states and resulting actions. These models are used to capture knowledge about the system behaviour and can be translated into other models (for example Markov models as discussed later in the report) to support further quantified analysis.

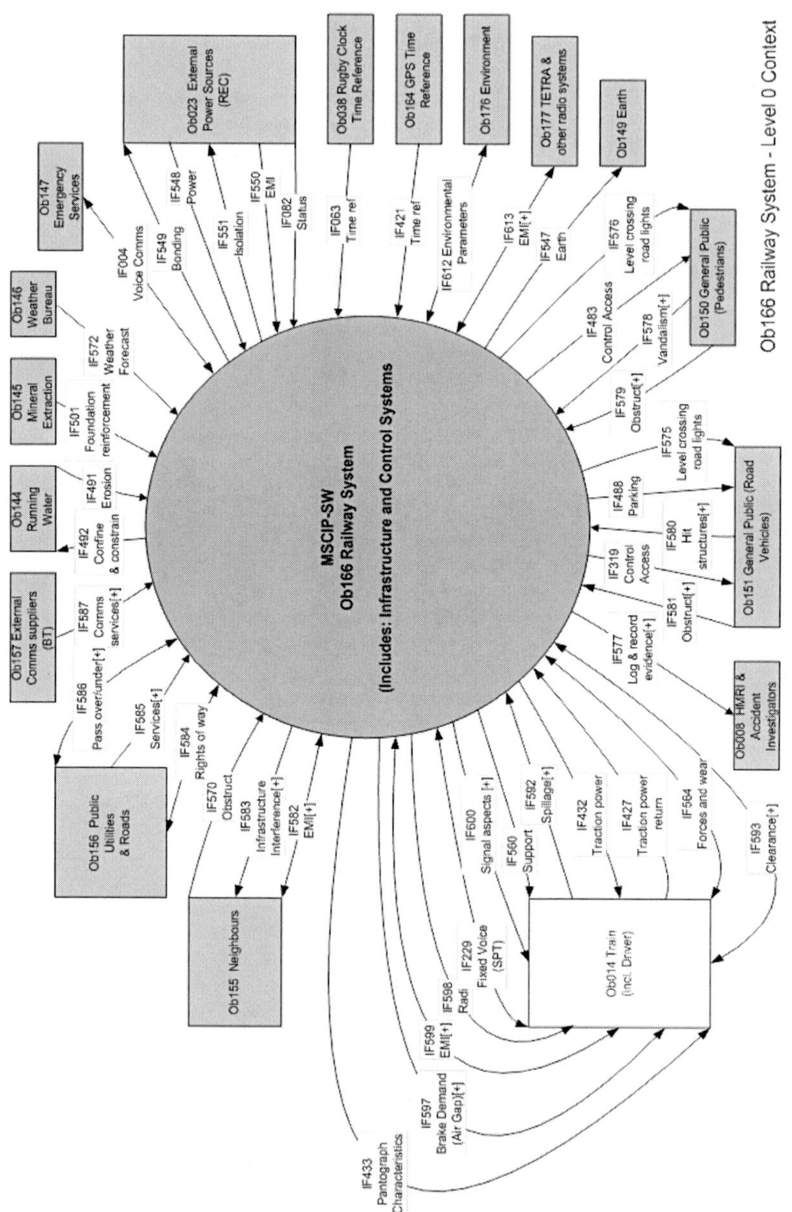


Figure 3-3: An illustrative example of the system decomposition model.

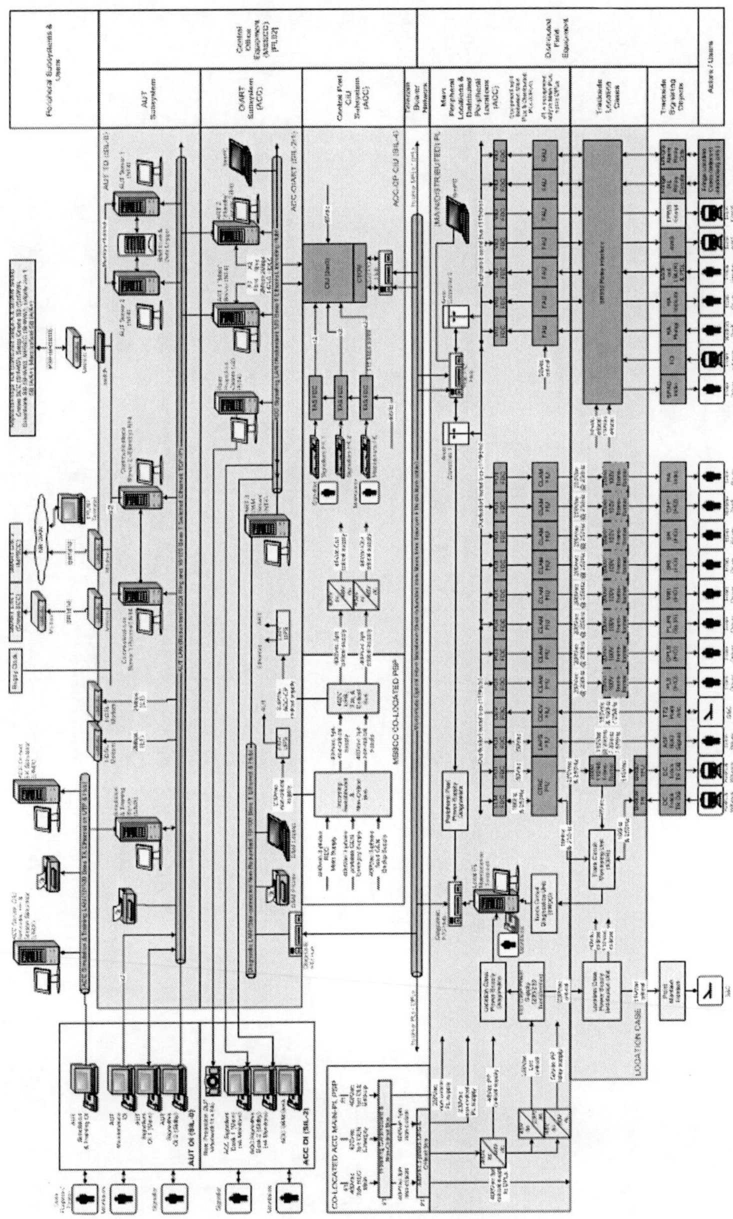

Figure 3-4: An illustrative example of the system decomposition model.

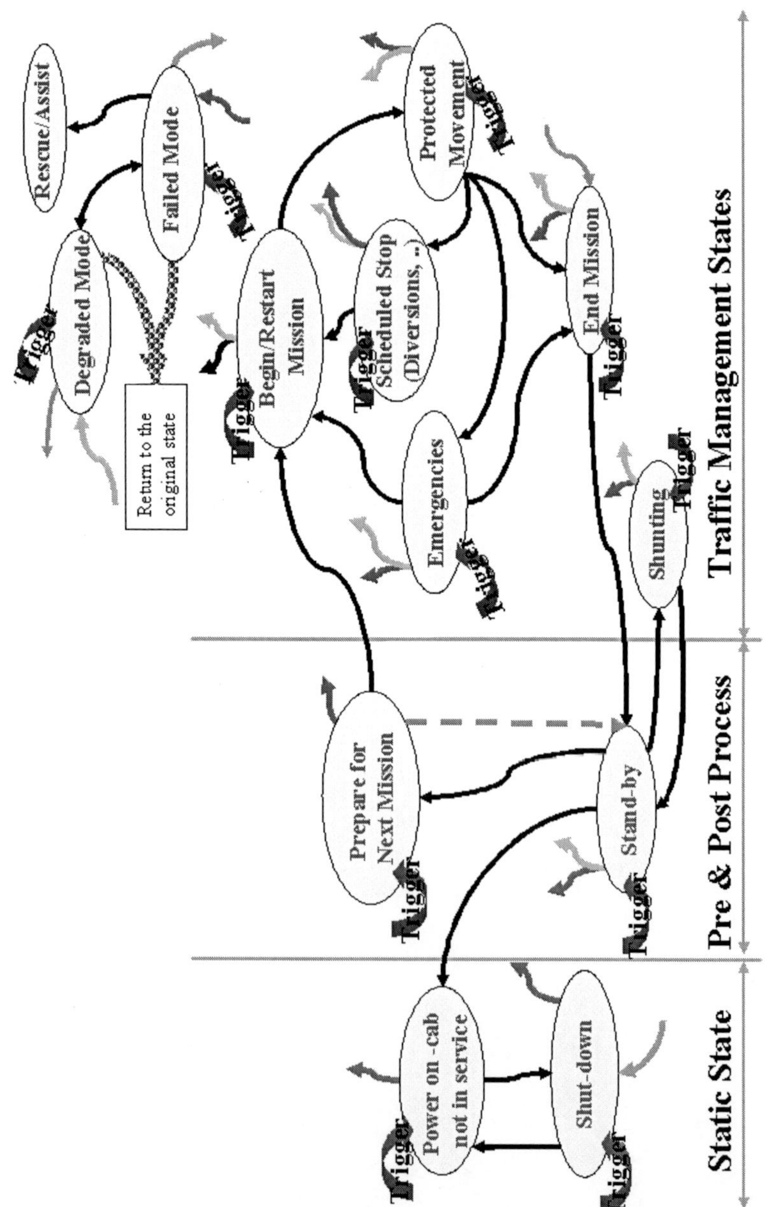


Figure 3-5: State transition diagram of a railway system from a train point of view.

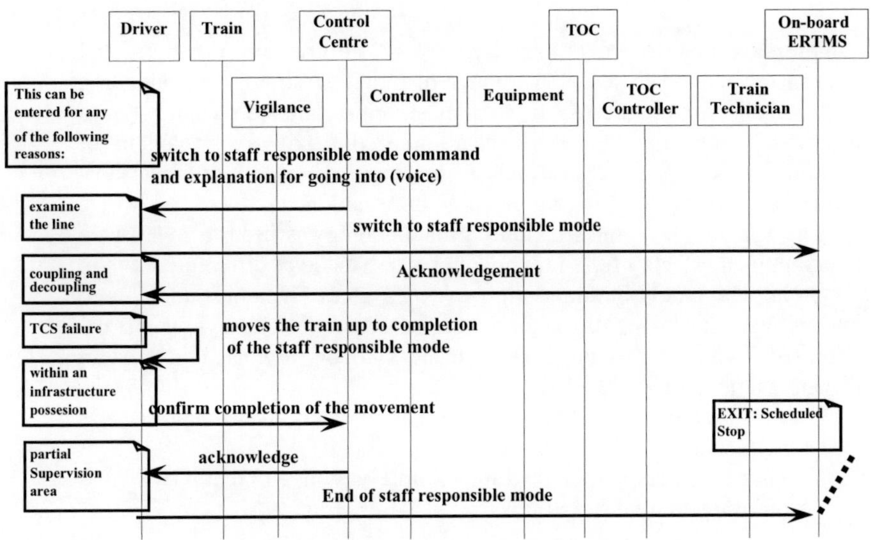


Figure 3-6: An illustrative example of sequence & collaboration diagrams

Sequence and collaboration diagrams (models) are interaction models that capture the dynamic nature of energy, information or action exchanges between system constituents or a system and the environment. For each state of the state transition model, a sequence and collaboration model can be produced to represent the system interactions for each system state.

System constituents are represented by shapes (usually rectangles) from which a vertical line (representing the existence of the relevant shape in time) is drawn downwards to capture the time and directed lines (arrows) representing the exchanges between system constituents. Shapes (usually folded corner rectangles) represent the entry and exit points.

These models are useful in support of system conceptualisation, definition, and analysis, as they are capable of encapsulating information related to temporal and spatial aspects of the system behaviour.

This technique has been used by the author of this report in support of a large scale project aiming to define safety requirements for European Railways Traffic Management System (ERTMS).

### 3.6.2 System Dynamics

System Dynamics is defined (Topintzi, 2001), (Sterman, 2000) as a rigorous method for qualitative description, exploration and analysis of complex systems in terms of their processes, information, organisational boundaries and strategies; it facilitates quantitative simulation modelling and analysis for the design of system structure and behaviour. It is a quantitative technique for the evaluation of the dynamic behaviour of socio-technical systems, and their response to the stream of decisions created in response to changing inputs of information.

The Systems Dynamics modelling methodology is used to investigate the combined effects of the individual changes made at different points in a system. The model-building approach of System Dynamics attempts to include and quantify all factors that influence the behaviour of the observed system or its parts. Two basic symbols are used in System Dynamics modelling:

1. Accumulation of something (information, energy, etc), represented by rectangle;
2. Flow or movement of anything (people, action, etc) that is represented by directed lines.

Together, these basic elements form cause and effect loops.

A vital principle of System Dynamics is to incorporate all information believed to significantly influence behaviour into the model, leaving out unnecessary detail. System Dynamics is also problem-centred, or goal-centred.

Together, accumulations and flows form cause-and-effect loops that model problem development over time and identify the likely consequences of the system.

The Systems Dynamics methodology was developed by the Massachusetts Institute of Technology professor Jay W. Forrester in 1961.

As a methodology for analysis (and management) of complex feedback systems, it has been used to address, practically, many different kinds of feedback system. System Dynamics caters for the description of the behaviour over continuous time and the identification of a system boundary to include the areas of interest for research and analysis.

It provides a structured format for identifying and depicting feedback system elements, goals, discrepancies, levels, rates and flows of information and controls.

The author of the book did not use this methodology in practice for following reasons:

1. System Dynamics modelling as a technique is suitable for high level abstraction. However, models required to support systems analysis on railways must be capable of supporting relatively detailed, low levels of abstraction;

2. It appears that in cases of more complex System Dynamics models, and in the case of railways certainly, the system models are very complex, it is practically impossible to calibrate these models.

### 3.6.3 Risk Matrix (Rapid Ranking)

Frequency-severity risk matrices are often used in support of risk analysis. They are matrices of likelihood and consequence categories. Individual risk values of cells of the matrix are assigned using qualitative analysis. However, risk matrices are only appropriate for the prioritisation of risks that are to be further analysed or acted upon.

The Risk matrices are deficient for the following reasons:

1. Subjective and without auditable justification;
2. Coarse, usually leading to overestimation of risk;
3. Based on extrapolated judgmental assessment of frequency of safety consequence arising from a hazard, missing the consequence analysis and, therefore, not covering the intricacies of all risk mitigation measures nor including the related consequences of other hazards.

### 3.6.4 Failure Mode and Effects (and Criticality) Analysis

Failure Mode and Effects (and Criticality) Analysis (FMEA/FMECA) are structured methodologies for the identification and analysis of the effects of latent equipment failure modes on system performance. This is a bottom-up process starting with the failure of a constituent/subsystem and investigating the effect of this on the system. It should be conducted by a team of experts with cross-functional knowledge of the analysed system, process or product. The methodology consists of the following steps:

1. Identify the object of the analysis (part, subsystem, step in the process, etc);
2. Identify the function related to the object of analysis;
3. Identify the failures of the object of the analysis;
4. Identify the effects of different failures on the object of analysis and the system which the object is part of;
5. Identify the causes of the failures;
6. Identify and analyse existing controls in place to mitigate the failures;
7. Identify any additional potential controls;
8. Recommend corrective actions to minimise the risk;

9. Prioritise implementation of corrective actions using a consistent standard.

The criticality analysis includes the estimation of probability of occurrence and severity.

### 3.6.5 Theories of probability

The two probabilistic theories that will be mentioned in this book are conditional probability and Bayesian Belief Networks (BBN).

Conditional probability theory was developed a long time ago, but is still one of the most useful tools for the analysis of uncertainty. Let us consider two variables, A and B, with the associated sets of their possible values ($a_1, a_2, a_3, \ldots, a_n$) and ($b_1, b_2, b_3, \ldots, b_n$).

Now if we know the conditional probability of $P(a_i \mid b_j)$, whenever the event $B=b_n$ is observed, the probability value of unobserved event $A=a_m$ can be calculated.

Bayesian Belief Networks are directed acyclic graphs. The nodes represent uncertain variables, and the edges are the causal or influential links between the variables. Associated with each node is a set of conditional probability values that model the uncertain relationship between the node and its parents.

The theory of BBN combines the Bayesian notion of conditional independence and probability theory. Once a BBN (Fenton, Neil and Forey, 2001) is built, it can be executed using an appropriate propagation algorithm.

This involves calculating the joint probability table for the model (probability of all combined stages for all nodes) by exploiting the BBNs conditional probability structure to reduce the computational space:

$$\underset{p(\lambda|e)}{\underset{distribution}{posterior\_failure\_rate\_}} = \frac{\underset{p(e|\lambda)}{\underset{given\_the\_failure}{likelihood\_of\_evidence\_}} \quad \underset{p(\lambda)}{\underset{the\_failure\_rate}{prior\_belief\_about\_}}}{\underset{evidence}{p(e)}}$$

Equation 1: BBN

Following diagram is a graphical representation of a BBN.

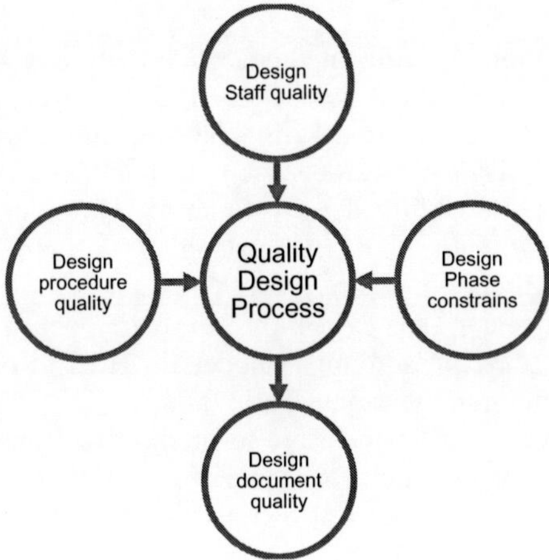


Figure 3-7: BBN representation

Once a BBN has been compiled, it can be executed and will exhibit the following two key features:

1. The effects of observations entered into one or more nodes can be propagated throughout the net, in any direction, and the marginal distribution of all nodes updated;
2. Only relevant inferences can be made in the BBN. The BBN uses the conditional dependency structure and the current knowledge base to determine which inferences are valid.

### 3.6.6 Monte Carlo Simulation models

Monte Carlo methods are statistical simulation methods that use sequences of random numbers to perform the simulation. With these methods, a large model of a system is sampled in a number of random parametric configurations, and the resulting data is used to describe the emerging properties of a system as a whole.

As opposed to a conventional numerical solution approach, which would start with the derivation of differential equations describing the mathematical model of the physical system, and then discretising the differential equations to solve a set of algebraic equations for the unknown state of the system, the Monte Carlo methods use random sampling techniques to arrive at a solution of the physical problem. Sometimes it is necessary to use other algebraic methods to manipulate the outcomes of Monte Carlo simulations.

When using Monte Carlo methods, the real, physical process is simulated directly, and there is no need for any derivation of the differential equations in order to describe the behaviour of the system.

The only prerequisite is that the system, or a model of the system, can be described by a probability density function. However, obtaining the accurate probability values and understanding the probability density functions can be difficult.

Sometimes it may be necessary to carry out supplementary experiments, or to develop additional models in order to obtain sufficient information for input into the Monte Carlo model.

In many applications, it is possible to predict the statistical error (the "variance") of the result, and consequently, it is therefore possible to estimate the number of Monte Carlo trials that are needed to achieve a given error interval.

### 3.6.7 Theory of Evidence

In 1976, Shafer (Bhatnagar and Kanal, 1991) developed the theory of evidence.

Probability theory states that if "S" is the universal set of events, then the uncertainty information consists of a probability density function, "p" such that

$$p: S \rightarrow [0, 1] \text{ and } \sum_{s \in S} p(s) = 1$$

Equation 2: Theory of evidence

In the theory of evidence, a mass function is used. The mass function is induced by the available evidence, and it assigns parts of a finite amount of belief to a subset of "S". Each assignment of a mass to a subset "s" of "S" represents that part of our belief that supports s without being able to allocate this belief among strict subsets of s.

The degree of belief in a subset "A" of "S", "bel(A)", is defined as the sum of all masses that support either "A" or any if its strict subsets:

$$bel(A) = \sum_{X \in A, X \neq \emptyset} m(X)$$

Equation 3: *Degree of belief*

The degree of plausibility of a subset "A" of "S", "Pl(A)" is defined as

$$Pl(A) = \sum_{X \cap A \neq \emptyset} m(X)$$

Equation 4: *Degree of plausibility*

or the sum of all the masses that can possibly support "A". The possibility of a subset "s" supporting "A" means that the mass assigned to "s" can possibly gravitate to that strict subset of "s", which is also a subset of "A".

### 3.6.8 Reliability Block Diagrams and Network models

Reliability Block Diagrams (RBD) are used to show a highly abstract (Pukite and Pukite, 1998) view of the system redundancy. The RBDs are simple, well known, and easy to evaluate. They use blocks to represent system constituents (Goble, 1998).

The blocks are arranged to correspond to the constituents required for successful operation. If there is a path through the block diagram with all the blocks being successful, this constitutes the operation being successful.

However, only two states can be represented for each constituent by RBD. This is a serious limitation since more often than not it is necessary to consider more than two states of each system constituent as part of the analysis. Network models are very simple models consisting of nodes and communication links. They are used to represent communication networks with individual links.

### 3.6.9 Fault Trees - Event Trees

Fault Trees and Event Trees are formalized diagrammatical models that trace paths from a top Hazard (unwanted event) to the basic input events via logic gates and the development of a Hazard through to the consequences resulting from success or failure of systems intended to mitigate effects of the initiating event.

A starting point of the Fault-Event Tree (FET) analysis is always the Hazard (Goble, 1998), (Hessami, 1999a), (RSSB, 2007).

These models can be used qualitatively, to capture and represent the scenario that gives a rise to a hazard and subsequently to its consequences, thus supporting options analysis and identification of new potential mitigations.

Moreover these models are used to quantify the frequency/probability of occurrence of a hazard and its consequences, thus potentially supporting impact assessment. An example of the Fault –Event Tree is presented in Figure 3-8 below.

Although the Fault –Event Tree models are used frequently, these modelling technique has some serious limitations:

1. The FET analysis is capable only of presenting a snapshot in time. Therefore if different system configurations exist during different phases of a mission, this technique cannot be readily used;
2. The FET analysis does not include loss analysis, therefore it does not support the complete risk assessment as such, but only a part of it;
3. Temporal and spatial aspects of the system behaviour cannot be modelled by this technique.

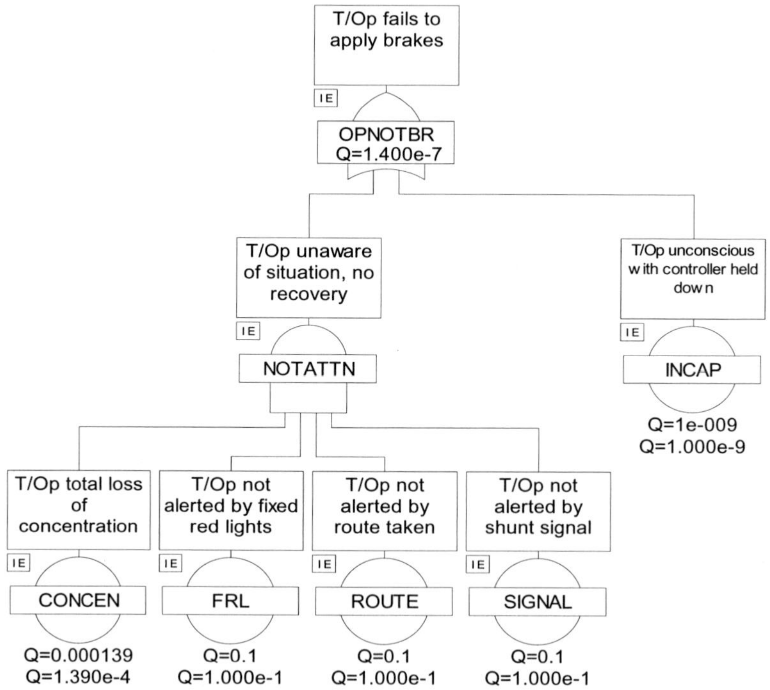

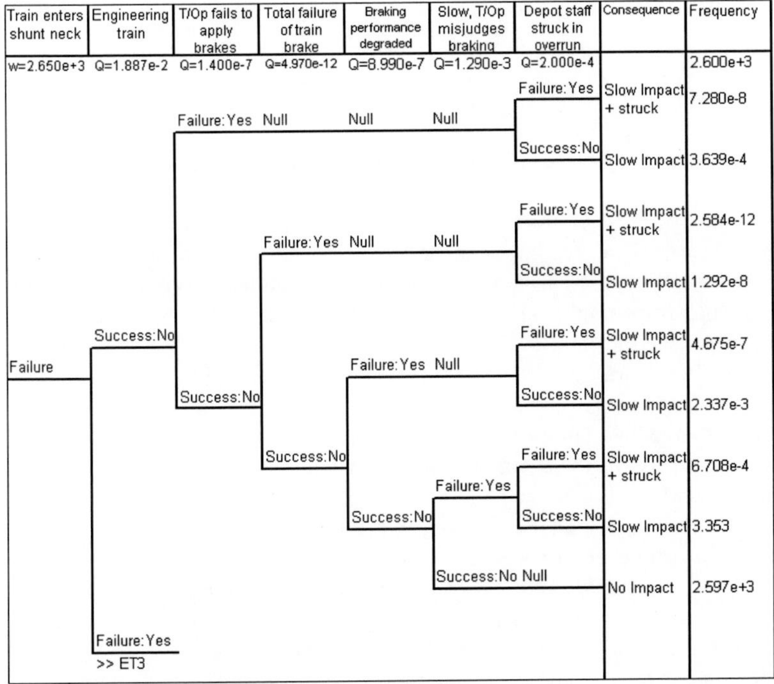


Figure 3-8: An example of Fault and Event Tree

### 3.6.10 Markov models

Markov models use state transition diagrams as their base. This technique uses only two symbols (Rouvroye, 2001); a circle to represent the successful/failed system constituents and a directed arc to represent possible constituent failure and repair.

The Markov models consider system states and possible transitions between them (Pukite and Pukite, 1998). The basic assumption in a Markov model is that the system has no memory, and that consequently, transitional probabilities are defined purely by the present state, and not its history. Probabilities of operation are calculated in each state as a function of time, taking into consideration the failure and repair rates.

## 3.6.11 Petri Nets

Petri nets are a graphical and mathematical modelling tool (Pukite and Pukite, 1998) and (Topintzi, 2001).

The basic objects of Petri nets are *places (represented by circles), transitions (represented by bars),* and the *arcs* that connect them. *Input arcs* enter transitions from places, and *output arcs* leave transitions and enter places. More symbols can be used. Places represent discrete states whilst transitions are active constituents. They model activities which can occur (the transition *fires*), thus changing the state of the system (the marking of the Petri net).

Transitions are only allowed to fire if they are *enabled*, which means that all the preconditions for the activity must be fulfilled (there are enough tokens available in the input places). When the transition fires, it removes tokens from its input places and adds some tokens at all of its output places.

As a graphical tool, Petri nets capture and represent the dynamic and concurrent properties of system. It is possible to set up state equations, algebraic equations, and other mathematical models governing the behaviour of systems. Petri nets can model a variety of situations and are easy to understand. However, the underlying model is difficult to solve. A larger model may become very complex and a solution may require the use of Monte Carlo tools.

## 3.6.12 Theory of Fuzzy Sets

Fuzzy logic simulates the human decision-making process, with the ability to generate precise solutions from certain or approximate information. While most of the other conceptions require an accurate mathematical model to simulate real-world behaviours, fuzzy logic is capable of using real-world human logic and even language.

Using fuzzy logic reduces the time required to interview and interpret a domain expert's knowledge of risk sources and consequences, while still capturing any uncertainty in the expert's opinion. Fuzzy logic is an alternative way to simulate uncertainty. Fuzzy number sets have a range, the wider the range the more uncertainty is introduced into the system. Fuzzy numbers have a degree of membership between zero and one associated with each point in the range.

The degree of membership is not a probability density function. The mean of the degree of membership curve can be found mathematically as the centroid of an area defined by the curve. The degree of membership curve does not need to integrate to 1, and integration of the curve does not produce a cumulative density function. There is no probability associated with the extremes of the range or any point within the range. Monte Carlo simulations are not needed for fuzzy numbers.

Fuzzy logic is a superset of Boolean logic, which has been expanded to accommodate the concept of the degree of truth-values, between the "absolutely true" and "absolutely false". The concept allows for mathematical modelling of the modes of approximate reasoning. The essential characteristics of fuzzy logic are:

1. Exact reasoning is perceived as a limiting case of approximate reasoning;
   Everything is a matter of degree;
2. Any logical system can be fuzzified;
3. Knowledge is interpreted as a collection of elastic or, equivalently, fuzzy constraints on a collection of variables;
4. Inference is viewed as a process of propagation of elastic constraints.

All the decisions that humans make are based on rules; if-then statements. Rules relate ideas and link one event to another, establishing cause-consequence relationships. Fuzzy logic mimics these decision-making processes; a decision, and the means of choosing that decision, are replaced by fuzzy sets, and the rules are replaced by fuzzy rules. Fuzzy rules operate using a series of if-then statements. Each fuzzy rule defines a "patch", a segment, of the decision-making process curve. A finite number of patches can cover a complete curve, defining the complete "decision" curve. If patches are large, then the rules are loose, if patches are small then the rules are refined.

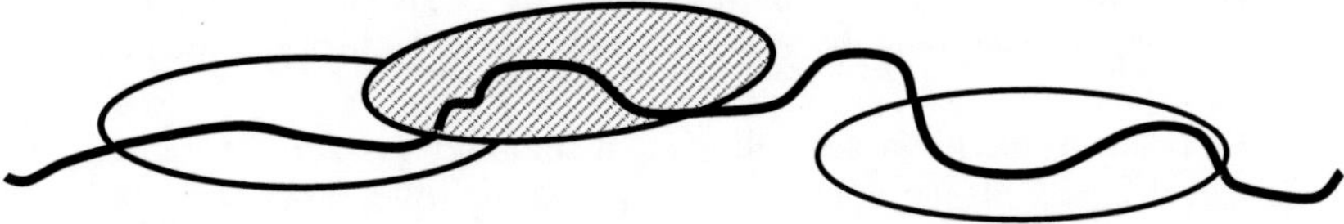

Figure 3-9: Fuzzy Patches

Fuzzy processors use expert knowledge which is expressed as a set of rules instead of, for example, differential equations to describe the system. A set of rules can be defined based on the expert knowledge using linguistic variables, which are described by a fuzzy set.

A linguistic variable is a quintuple (X, T(X),U,G,M), where “X” is the name of variable, T(X) is the term set, set of names of linguistic values of “X”, “U” is the universe of discourse, “G” is the grammar to generate the names and “M” is a set of semantic rules for associating each “X” with its meaning.

Three steps are taken to create a fuzzy process:

1. Fuzzification (Transformation of measurements/crisp values into fuzzy values);
2. Rule evaluation (Application of fuzzy rules, also known as inference);
3. Defuzzification (Obtaining crisp values or actual results).

The employment of Fuzzy Processors is recommended for very complex processes, when there is no simple mathematical model, for highly nonlinear processes, and if the processing of linguistically formulated knowledge is to be performed.

If we define a process surface as a hyper-plane derived from a multiple set of process inputs/output relationships, it will be possible to relate inputs to outputs, and tune the processor by altering the rulebase and comparing the effect on the process surface. Each point on the plane has its coordinate, [x y z], defining the position within the envelope relating inputs, [x y] to output [z]. This means that the fuzzy processor can be tuned by shaping the process surface, rather than by adjusting numerical gains.

Furthermore, it allows for complex, multiple process goals, and for nonlinear systems to be defined by relatively simple distortions of the process surface shape. Each rule in the rulebase defines a separate processor goal. Changes to individual rules do not have a global effect on the process surface, thus the processor action can be locally tuned to suit specific need.

Overlaps of the fuzzy sets, allowing a smooth transition between rules, define an area of the input range where multiple input sets can be activated. This area is known as the Area of Influence. Increase of the overlap level, between the fuzzy sets, increases the proportion of the process surface contained within the Area of Influence.

Since the de-fuzzified crisp value of the process output is located within the Area of Influence, an increase of the proportion of the process surface contained within the Area of Influence will smooth the processor action. In order to smooth the process action, but keep the ability to distinguish between different regions of the process surface at the same time, an overlap of the fuzzy sets should not extend over more than 50%.

In order to reach a compromise between the smoothness of process and resolution of surface, it is possible to use more advanced sets of shapes as membership functions, such as bell shaped membership functions instead of simple triangular shaped ones, to describe the fuzzy set.

### 3.6.13 Weighted Factors Analysis

Weighted Factors Analysis (WeFA) methodology is a creative knowledge capture, representation and evaluation methodology (Lucic, 2004a), (Hessami, 1999a), (Hessami, 1999b) and (Hessami and Hunter, 2002). It enables the knowledge about the given problem to be captured and represented at a high and strategic level. WeFA methodology supports a holistic approach to problem solving.

In contrast to a risk based perspective focusing on hazards alone, WeFA analysis also highlights the opportunities that require effort at enhancement and optimisation alongside reduction in risks. The knowledge elicitation process is group based, exploiting the diversity of expertise and perspectives. This first stage of the analysis is focused around a task identification of an AIM. The AIM is a common goal/objective of the undertaking, or of a system being analysed.

Once an AIM is defined and agreed (for example the AIM may be "Enhancements of the system performance") the group is encouraged to identify the highest level DRIVERS and INHIBITORS that are likely to influence the AIM. The DRIVERS are defined as factors that contribute to the AIM and the INHIBITORS as factors that are detrimental to the AIM.

An example is presented by Figure 3–10 and Figure 3-11 below. Let us assume that the study seeks to analyse factors influencing safe driving. So, the agreed aim is "Safe Driving". Furthermore, let us assume that the workshop participants concluded that clarity of rules and procedures and good, comprehensive training enhance the ability of the driver to drive safely. These two will be the DRIVERS. In the same fashion, let us assume that the workshop participants concluded that bad weather conditions could hamper driver's attempts to drive safely. This is the INHIBITORS in the WeFA terminology.

The elicitation process should continue for each DRIVER and INHIBITOR depending on the need to understand its properties. The elicitation is terminated within the branch when the group feels sufficient clarity has emerged and further decomposition is not likely to add value. Figure 3-11 below, illustrates this approach.

WeFA underpins following principles:

1. Definition and group argument on the focus of the analysis;

2. Consideration of inherent polar opposites as influencing factors;
3. Hierarchical and successive decomposition;
4. Consideration and inclusion of hard and soft factors into the analysis;
5. Simple graphical representation of knowledge;
6. Weighting of factors according to their degree of influence;
7. Explicit representation of dependency between factors;
8. Potential for quantification and treatment of uncertainty;
9. Problem resolution through options identification and evaluation.

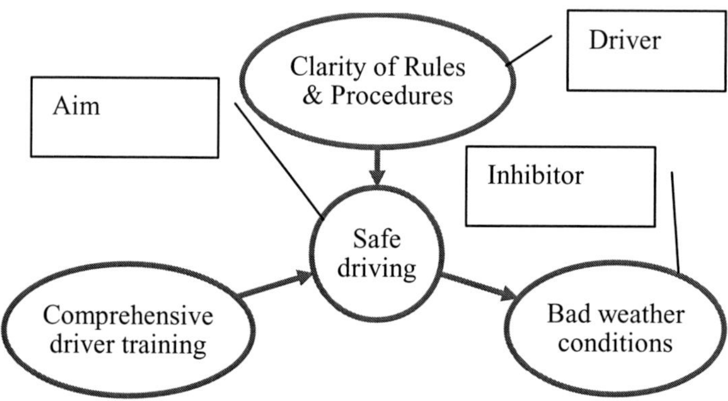


Figure 3-10: WeFA example

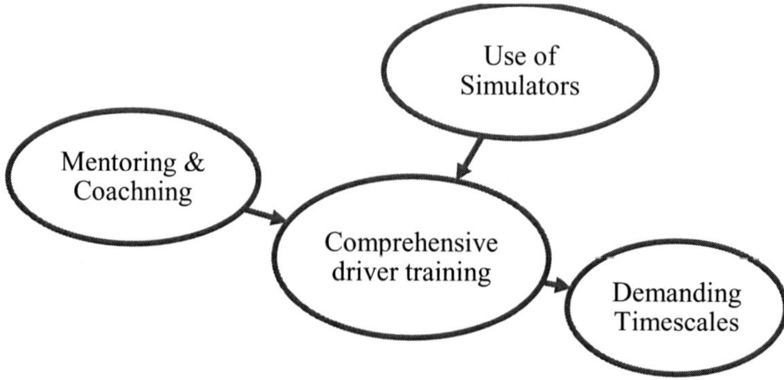


Figure 3-11: WeFA example (expanded)

The author of this book used the WeFA methodology in support of data elicitation for a strategic railway project, and it was observed that the WeFA methodology is especially beneficial when used in support of high level, strategic knowledge elicitation (Lucic, 2001).

Although several different technologies have been tried, (BBN, Fuzzy) at the moment, the general mathematical theory for quantified WeFA analysis is not available.

### 3.6.14 Advanced Cause – Consequence models

Once the significant hazards within the scope of a product, process or undertaking are identified and recorded, it is desirable and constructive to understand their causality.

This may help in eliminating the hazardous condition altogether, or to take actions to reduce its likelihood or frequency. The process is referred to as Causal Analysis. The Modelling methodology used in support of Causal Analysis is commonly known as Fault Tree Analysis.

Irrespective of the extent of preventative measures taken, hazards do occur. It is also desirable to develop an understanding as to how a given hazard is likely to escalate in the real world once it has occurred, and what range of incidents and accidents are likely to arise from the hazard.

The idea being that even in the occurrence of a hazard, a range of measures may be available to detect, control and avoid accidents, or minimise losses associated with such accidents. This process is referred to as Consequence Analysis. Consequence Analysis involves establishing the intermediate conditions and final consequences that may arise from a hazard.

It involves a bottom-up assessment of each hazard and is focused on the post hazard horizon. The sequence of intermediate conditions identified is termed "the hazard development scenario". Defences against potential escalation (the defence may be equipment, procedure or circumstance) are referred to as 'Barriers'. The Consequences of a hazard are categorised into three broad categories, Safety, Commercial and Environmental. An example of basic Cause- Consequence risk model structure is given in Figure 3-12.

The following are objects that are defined within the framework of the advanced cause consequence model:

1. Base Events represent different failure modes and events that may trigger the hazard and are related to failures of system constituents. Base events are numerically expressed as failure rates or probabilities.

2. Logical gates "AND" and "OR" are used to mimic possible logical combinations of different failures and therefore describe the hazard evolution scenario. Together with base events they make up a "fault tree".
3. Hazard or Critical Event whose probability or frequency of occurrence is calculated by the fault tree.
4. Barriers are defences against a hazard. Once a hazard is already "live" usually there are still some defences, prevention measures which avert the hazard from becoming a consequence. Barriers are defined by the probability of success of the barrier working correctly, and therefore preventing the consequence. Therefore, barriers have one input, the probability of a hazard occurring, and two outputs, the probability of the barrier working (success) and that of it not working (failure) There are three different types of barriers:
   a. Physical Barrier; equipment based protection measure, for example fire alarm;
   b. Procedural Barrier: procedure based protection for example evacuation procedure;
   c. Circumstantial Barrier: if a consequence is avoided by 'pure luck'. It is beneficial to record these, as it may be possible to turn them, in future, into a physical or procedural barrier and therefore mitigate against the hazard better.
5. Consequences are defined as the final outcomes of the hazardous scenario, for example a collision between two trains. The Cause Consequence model calculates the probability or frequency of occurrence of a consequence. To evaluate the risk of a consequence, it is necessary to estimate a potential loss associated with each of the consequences. The estimation is not part of a cause consequence modelling tool. Consequences are grouped into 4 categories as follows:
   a. Predominantly Safety Related Consequence: If a predominant loss associated with the consequence is safety related, then the consequence is classified as a safety consequence;
   b. Predominantly Commercial Consequence: If a predominant loss associated with the consequence is commercial (damage to property, loss of service, etc.) then the consequence is classified as a commercial consequence;
   c. Predominantly Environmental Consequence: If a predominant loss associated with the consequence is environmental (damage to nature, release of contaminated materials, etc.)

then the consequence is classified as an environmental consequence;

d. Broadly Safe Consequence: if a hazard, occurs but due to Barriers working does not give a rise to a safety, commercial or environmental consequence, in other words if the accident is avoided, then this outcome of the hazard development scenario is defined as a Broadly Safe consequence;

6. Internal to model connectors, "OUT" and "IN": due to the complexity of some of the cause consequence models, the modelling technique allows for models to spread across several pages, "work sheets" and for outputs of model elements to be transferred across different "work sheets". This is realised through use of "OUT" and "IN" symbols. This is purely linking functionality and does not transform the transferred probability or frequency in any way.

Standard Fault Tree analysis rules apply, including a need to ensure that Base Events are independent from each other. It is possible that the same base event may lead to several different consequences, and hence appear in several places in the same model. In such cases it must be assumed that the base events and relevant data are described and justified within the specific context each time.

This can be achieved by means of the use of logical gates to combine these Base Events with other ones to capture the specific context.

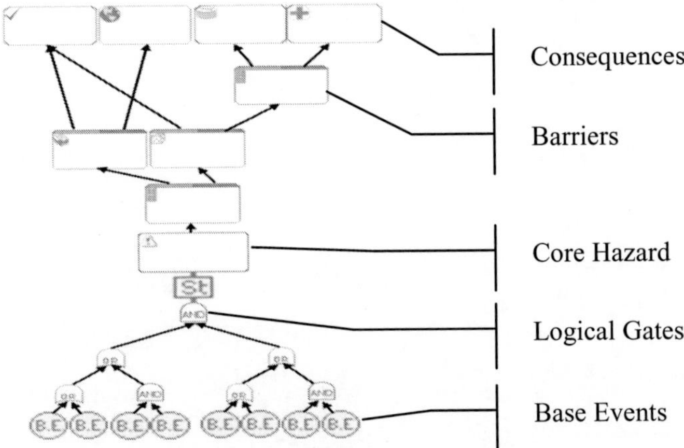


Figure 3-12: Basic structure of the Cause-Consequence Model

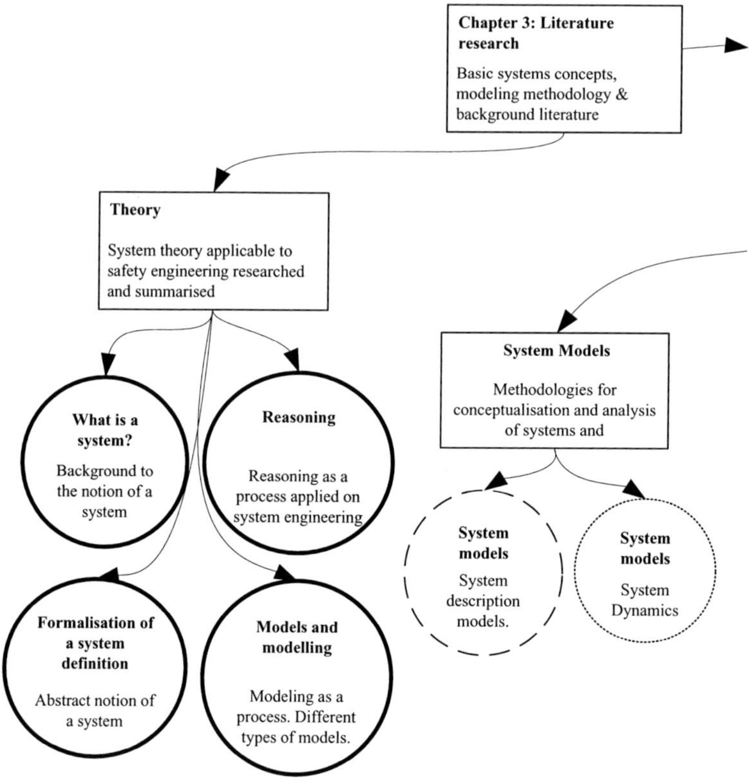

**Chapter 3: Literature research**

Basic systems concepts, modeling methodology & background literature

**Theory**

System theory applicable to safety engineering researched and summarised

**What is a system?**

Background to the notion of a system

**Reasoning**

Reasoning as a process applied on system engineering

**Formalisation of a system definition**

Abstract notion of a system

**Models and modelling**

Modeling as a process. Different types of models.

**System Models**

Methodologies for conceptualisation and analysis of systems and

**System models**

System description models.

**System models**

System Dynamics

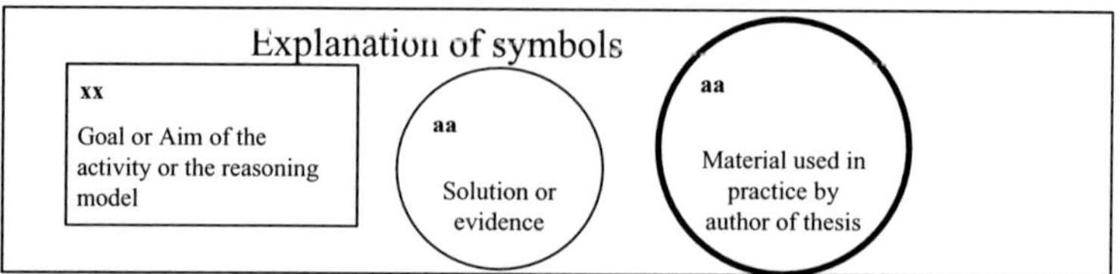

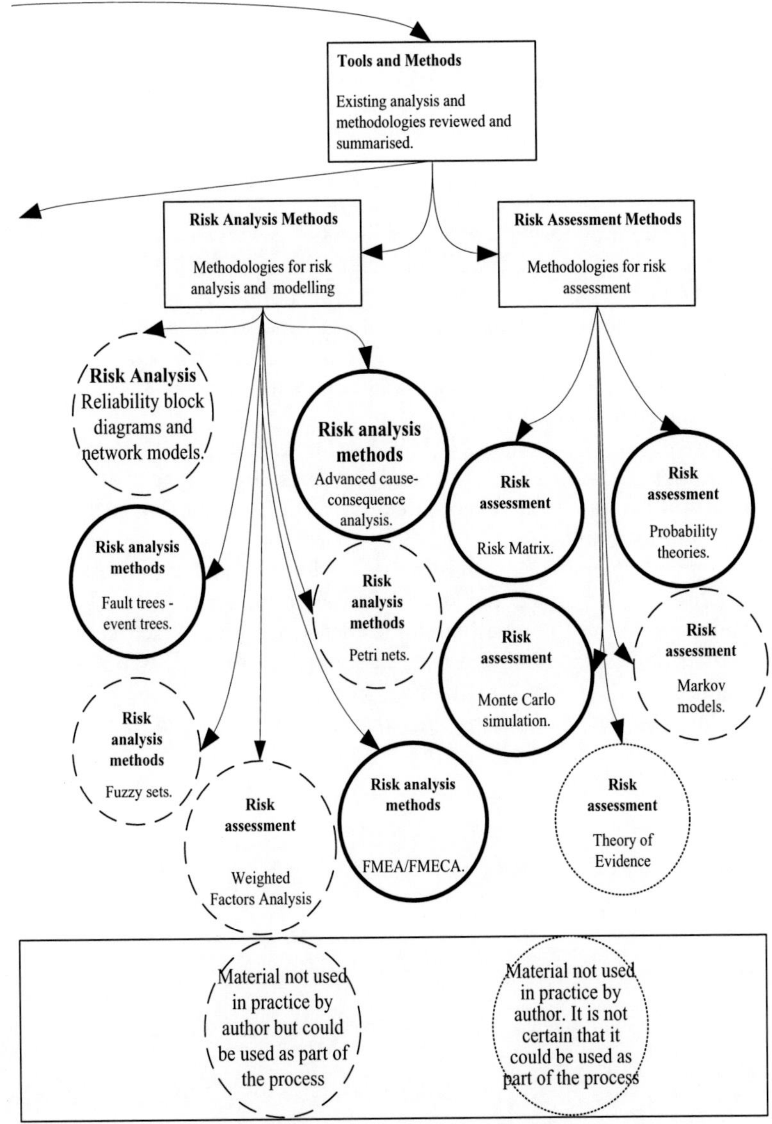


Figure 3-13: Overview of analysed methodologies and their applicability on the problem domain.

The Advance Cause Consequence Modelling technique has been further developed into a parametric modelling technique, and used on several railway projects. The most recent application of the technique is described in the following chapter of this book.

## 3.7 Chapter Conclusions

In this chapter, the findings of the literature review are presented. Two specific areas of interest were subject of the research. Firstly, system theory applicable to safety engineering, the notion of a system, formal definition of the system as a concept, the modelling process and the reasoning process were investigated and are summarised here. The knowledge gained from this activity has been used as an essential building block in the later development.

Following on from that, a review of the existing analysis tools and methodologies was carried out and findings from that research are summarised here as well. A more detailed critique of these is provided in chapter 5 of the book.

The main criticism of the existing methodologies is that they do not make a holistic, integrated and heterogeneous framework.

However, each of the methodologies described in this section is useful in its own right, and should be used in support of the analysis of specific problem areas. Figure 3-13 below provides an overview of all the methodologies reviewed here and their applicability on the problem domain.

# CHAPTER FOUR

# THE APPLICATION AREA AND INITIAL DEVELOPMENT

## 4.1 Chapter introduction

As already discussed in the previous section, the existing techniques for analysis and assessment of safety risk as an emerging property of the System do not amount to an integrated methodology for analysis and assessment of complex systems.

Some of the techniques (detailed earlier) the author used in practise, to support delivery of four different projects.

However, it was necessary to develop these techniques further, and in some cases develop new ones, and then integrate them into a coherent, consistent methodology.

Although these four projects provided a drive for further development and a novel approach, due to the limited scopes of the projects at that stage it was not necessary to develop a holistic integrated methodology for the Engineering Safety and Assurance Process (ESAP), mentioned earlier in the book, as an integrated framework. The Victoria Line Upgrade Programme (detailed later) was at an early stage in its lifecycle (sufficiently early in the project, at design stage) and provided both a drive and justification for the evolution of the process and the methodology. Using the experience and knowledge gathered through work on the four projects depicted in this section, the ESAP was developed.

As part of the research, these projects have been used to test some of the novel concepts and approaches.

The following is the description of the practice, and the processes that were followed. The novel application of the existing methodologies and the development of novel methodologies by the author are indicated appropriately.

The structure of this chapter is outlined by Figure 4-1 below.

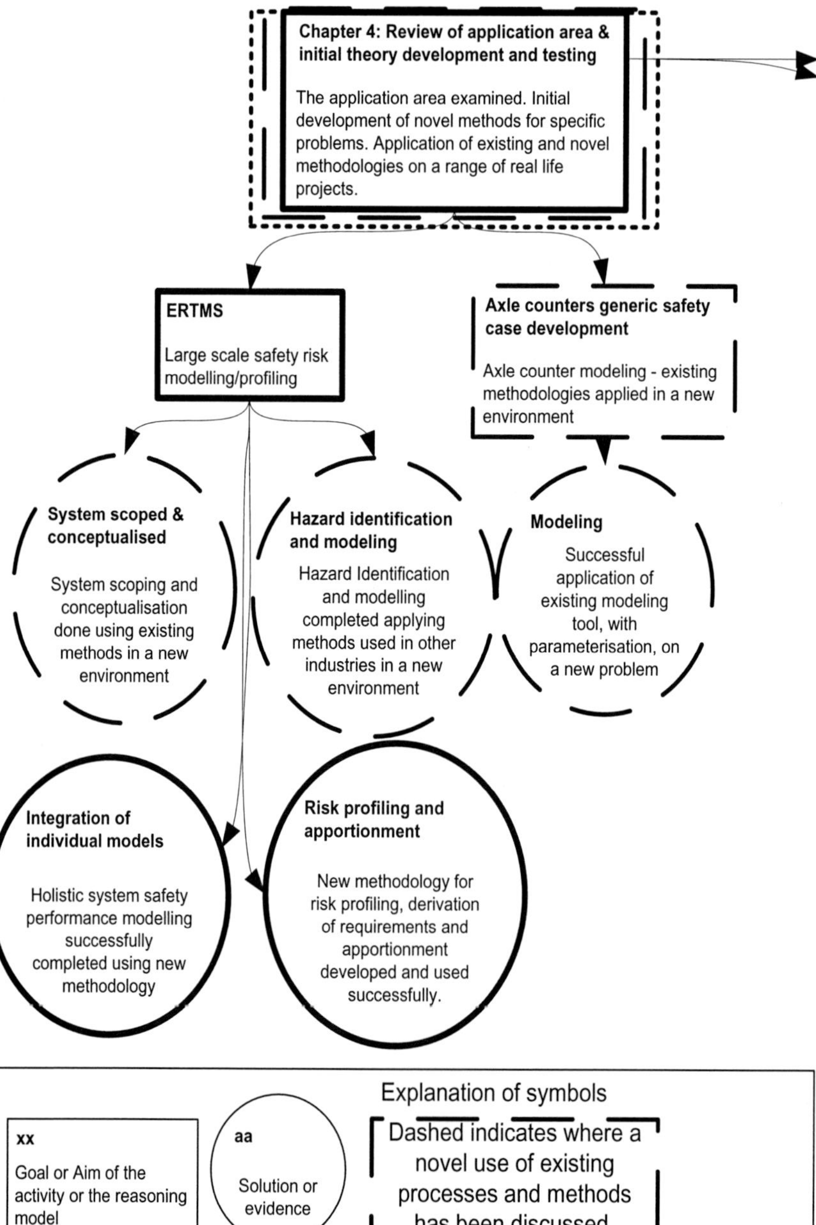

Chapter 4: Review of application area & initial theory development and testing
The application area examined. Initial development of novel methods for specific problems. Application of existing and novel methodologies on a range of real life projects.
ERTMS
Large scale safety risk modelling/profiling
Axle counters generic safety case development
Axle counter modeling - existing methodologies applied in a new environment
System scoped & conceptualised
System scoping and conceptualisation done using existing methods in a new environment
Hazard identification and modeling
Hazard Identification and modelling completed applying methods used in other industries in a new environment
Modeling
Successful application of existing modeling tool, with parameterisation, on a new problem
Integration of individual models
Holistic system safety performance modelling successfully completed using new methodology
Risk profiling and apportionment
New methodology for risk profiling, derivation of requirements and apportionment developed and used successfully.
Explanation of symbols
xx
Goal or Aim of the activity or the reasoning model
aa
Solution or evidence
Dashed indicates where a novel use of existing processes and methods has been discussed.

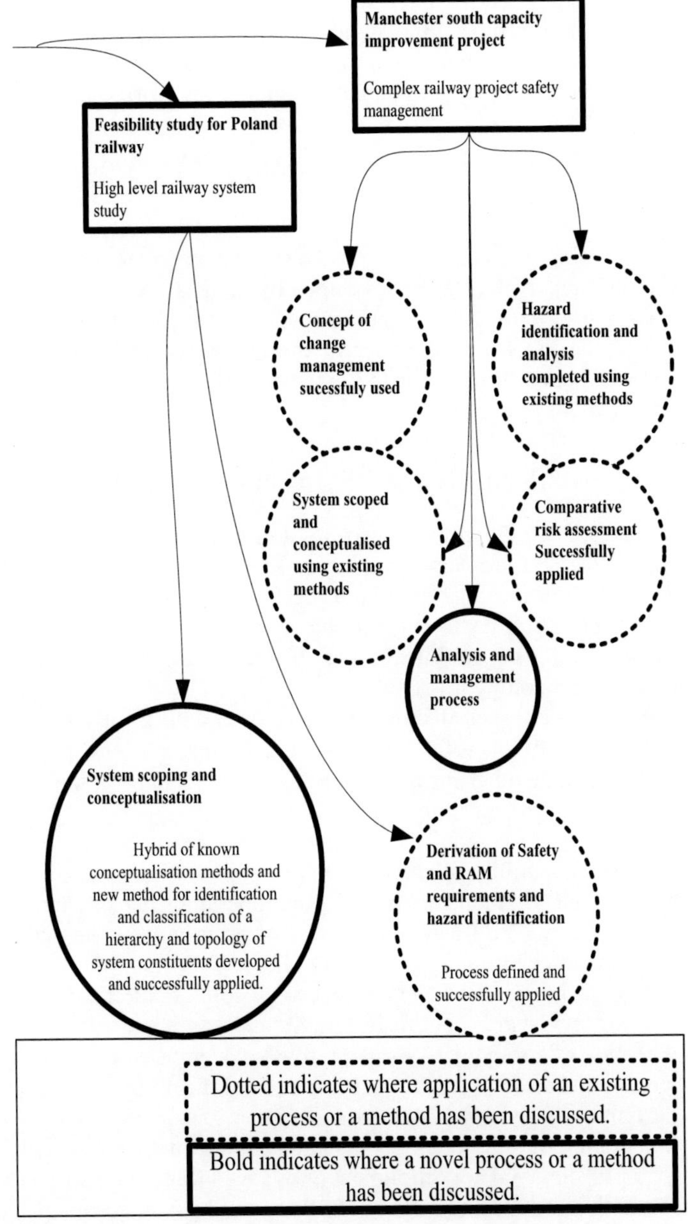


Figure 4-1: Outline structure and the argument presented in Chapter 4

For each of the projects, the author outlined the purpose and scope of the project, and presented the work and research carried out, the development of new techniques and the results of the work.

## 4.2 First Project: Large Scale Safety Risk Modelling – ERTMS

The modelling technique used on the project is novel. The technique and the tools supporting it were developed by the Risk Analysis Unit team of Railtrack Plc. The author was involved in the development of the process and the tools as a member of the team, and has developed the algorithms and specification for the extension of a tool in support of the modelling work as outlined later.

### 4.2.1 Purpose and Scope of the Project

The purpose of the project was to define and quantify the safety requirements for European Railways Traffic Management System (ERTMS). The European Commission for Transport managed the project, the biggest pan European engineering undertaking ever in terms of finances involved, geographical spread, required workforce and the complexity of technology involved.

The aim is to connect all European countries with a fully compatible high-speed rail network (500 km/hour). Later phases of the project are envisaged to include other former eastern European countries, the Balkans, Greece, Russia and Asian countries.

Standardised safety requirements and safety targets are necessary to achieve full compatibility between trains, railroads (track, signalling, power supply, telecommunications, etc.) and operating rules. A number of novel techniques, as well as extensions of existing techniques and tools, have been used to enable safety risk modelling as follows:

1. Use of state transition and sequence and collaboration models in support of Hazard Identification;
2. Creation of a reference, virtual, railway system used to derive a common data set to enable development of the unified risk model;
3. Extension of existing tools and techniques in support of the integration of individual models into a single integrated model for risk profiling and apportionment (Advanced Cause Consequence Modelling has been developed further into the Parametric Advanced Cause Consequence Modelling technique supported by a tool “Integrated Safety Assurance Environment (ISAE)”).

These developments are depicted in following sections.

### 4.2.2 Problem Description and Conceptualisation

The analysed system is not a uniform system; it is different in many aspects at different locations throughout Europe. At the time of the project, three different technical solutions, each with several variants, were planned to suit the diverse needs of national railways as well as the varied financial potential of various EU member states. Nevertheless, it was necessary to define the system to facilitate identification of the generic hazards resulting in a common, universal set of safety targets.

The following is a list of the problems the project has to resolve:

1. No detailed specification;
2. No single system solution;
3. No generic description of a system;
4. Different operational rules in different countries;
5. Different operational environments;
6. Different safety cultures and legal regimes throughout the EU;
7. No uniform data collection system;
8. Different stake-holders and expectations;
9. Different activities, tools and techniques;
10. Different degrees and measures of control;
11. Vastly different time-scales & costs;
12. Vastly different risk profiles;
13. Different people and competencies;
14. Different needs (freight versus passenger services, etc.);
15. Different levels of technological development and density of existing infrastructure.

However, initial analysis showed that the system was sufficiently known in generic terms that it could be broken down into elements that broadly fit into one of three categories:

1. Environment;
2. Constituents;
3. Operational modes.

Environment is defined as the environment within which the system works, including people, weather, climate conditions, traffic density, etc. Constituents are elements of the system, its hardware and software and people. Different operational modes are in fact dictated by different technological and procedural solutions to the same problem, namely:

Level 1 operational mode, or the railway system as it is now, but operated in accordance with new synchronised operational rules;
Level 2 operational mode, or 'in cab' signalling systems, where information to the train is transmitted via short distance trackside transmission antennas positioned at carefully chosen places so to mimic the functions of the conventional signal;
Level 3 operational mode or 'in cab' signalling where information to the train is transmitted via radio signal securing continuous and almost instantaneous information update.

Within any of the operational modes described above, a number of operational states, generic situations, were identified; and in such a way that one could say that at any moment in time, a train has to be in one and only one state or situation.

### 4.2.3 System representation and scoping

In support of the system representation and scoping, existing modelling techniques, "State Transition" and "Sequence and Collaboration" modelling were used in a novel way. A high level generic system description had been developed and used as the starting point of the analysis, using the state transition model. The focal point of this representation is a train ( Figure 3-5, earlier in the report). This, high level description, referred to as "ERTMS State Transition Diagram", depicts all the possible transitions between different states of a train within the ERTMS operational environment.

Three main classes of system behaviour have been identified:

1. Static States: The train does not move, none of the vital train management functions are engaged and the train is either shutting down or powering on;
2. Pre & Post Process: Preparation for the journey, arming of the computer or resetting the computer data and waiting for change towards "preparation for next mission";
3. Traffic Management States: The actual journey, including all possible states of a normal journey as well as emergencies and degraded operations.

In total 13 states have been identified:

1. Shut Down;
2. Prepare For Next Mission;
3. Scheduled Stop;
4. Emergencies;
5. End Mission.
6. Protected Movement;

7. Power On-Cab Not in Service;
8. Degraded Mode;
9. Rescue / Assist
10. Begin/Restart Mission;
11. Stand-By;
12. Failed Mode.
13. Shunting;

For each identified state, a so called pseudo-Spec has been developed, depicting the state itself and transitions from one state to another including the description of the transition trigger.

Subsequently, the "ERTMS State Transition Diagram" has been further analysed, and each of the individual states has been described in detail using "Sequence and Collaboration" analysis (example in Figure 4.2).

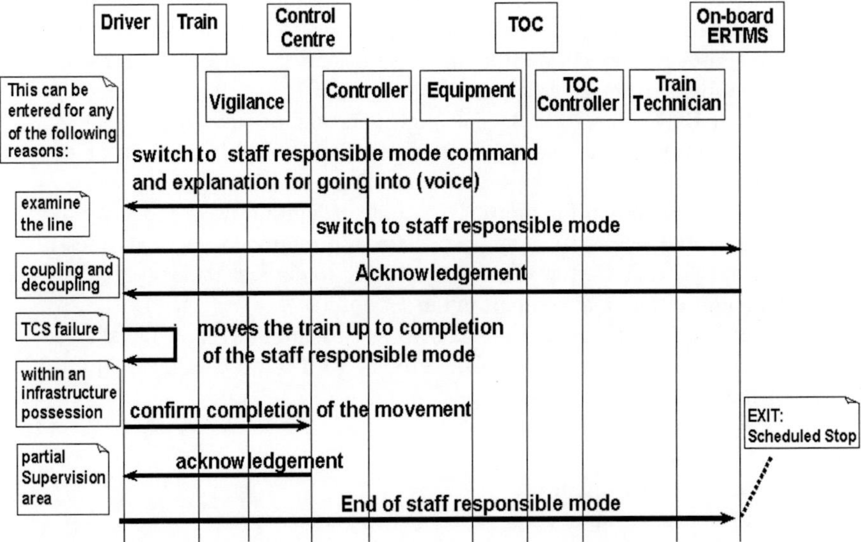


Figure 4-2: Authorised Movement - Staff Responsible [Sequence & Collaboration diagram]

All the actors/constituents of the system taking part in a state's existence are positioned on the top of the diagram. The vertical axis represents time, and horizontal directional lines define exchanges between the actors.

These exchanges can be communication or command/control exchanges, and are distinguished by the use of full and dotted lines respectively. Transition points (entry to the state and exit from the state, including transition triggers) have been described as well.

These representations have been used as the base of the hazard identification process.

Each of these exchanges has been analysed one by one.

### 4.2.4 Focus and Identification

***Hazard Identification***

Usually a Hazard Identification (HAZID) session is structured around loose system descriptions or system block diagrams. However, in support of the HAZID for this project, the system scope and context was depicted using the State Transition, Sequence and Collaboration (S&C) Modelling. This novel use of existing modelling techniques, in support of the hazard identification, resulted in a structured and systematic base for the hazard identification as the hazards resulting from each process or interaction were elicited (Lucic, 2005c). The consideration of keywords (such as NOT, LATE, EARLY, ALSO, etc) in relation to each exchange of the S&C models was used to support the HAZID process.

In addition, a novel function of the context diagram was to identify the experience and expertise required from the members of the hazard identification panel. Participants were allowed to express their opinion and encouraged to take an active part in the session.

In order to secure as comprehensive as possible a source of data and ideas, care has been taken to form groups with a wide range of trans-national expertise:

1. Engineers with design experience;
4. Operation managers;
5. Maintenance engineers;
6. Performance and Safety Performance managers;
7. Safety & Standards managers;
8. Train drivers.

Identified hazards were carefully recorded to be used as the basis for further analysis.

There are many benefits to "on-line" capture, using a computer and projector, the main one being that the panel can agree (or argue about, which is more usual) the final wording of the hazard and its consequences and mitigations. The form, designed in Microsoft ACCESS and projected on the wall during the HAZID sessions was used to capture all the elicited information.

***Analysis of hazards***

Around 150 hazards were identified. This is too many hazards to analyse individually, and furthermore, many are interrelated anyway. Consequently, all similar and/or related hazards were grouped into higher level groupings, clusters of hazards, labelled Core Hazards.

Core Hazards were the base cells of knowledge integrated into the knowledge structure. In a way, the Risk universe was partitioned into galaxies of hazards.

Initially, a "risk universe" with 10 "galaxies", Core Hazards, was formed:

1. Less Restrictive Movement Authority;
2. Driver Exceeds allowable Movement Authority;
3. Train Exceeds allowable Movement Authority;
4. Inappropriate Routing of Train;
5. Errors & Failures during Shunting;
6. Errors & Failures during Rescue;
7. Errors & Failures during Coupling/Joining;
8. Errors & Failures during Shutdown/Power up;
9. Errors & Failures during Standby/Prepare for Next Mission;
10. Inappropriate Level Transitions.

However, some of the Core Hazards were too complex to be modelled, and hence, some of them were divided into a Number of Sub-Core Hazards. At the end 19 Hazard Groupings were established. Following the analysis of each core hazard, 40 Cause – Consequence models had been developed, because for some of the core hazards, more than one model was needed in order to adequately model the safety risk.

### 4.2.5 Core Hazards Modelling

Causal Analysis and Modelling is the first part of the modelling process. The aim of causal analysis is to identify all the events/causes, and all the intermediate conditions that lead to a hazardous condition (Lucic, 2005d).

Each of the Core Hazards has been analysed in turn, attempting to determine all possible causalities of the Core Hazard, and the logical relationships between identified causalities leading to the materialisation of the Core Hazard. As result of this, a fault tree has been developed for each Core Hazard.

In a fashion similar to causal analysis, consequence analysis of each core hazard had been undertaken. Consequence analysis involves establishing the intermediate conditions and consequences that may arise from a hazard. It involves bottom-up assessment of each hazard and is focused on the post hazard horizon. At each intermediate state existing defences against potential escalation were identified. The defence may be equipment, procedure or circumstance.

The sequence of identified intermediate conditions is termed “the hazard development scenario”. The consequences fit into one of the following categories: predominantly safety related consequence, predominantly commercial consequence, predominantly environmental consequence, broadly safe condition. The combination of a causal model of a core hazard and the consequence model, of the same core hazard, results in the Core Hazard Cause-Consequence Model. The model calculates the frequencies or probabilities of occurrence for all the consequences within the model.

To estimate the risk that each of the consequences brings to the system it is necessary to calculate the loss that each consequence may give rise to. To estimate the loss for each of the consequences a software based tool was used, General Loss Estimation Engine (GLEE).

Loss for each consequence were characterised through the following categories:

1. Safety loss to Passengers:
    a. Minor Injury;
    b. Major Injury;
    c. Fatality;
    d. Equivalent fatalities;
2. Safety loss to Worker:
    a. Minor Injury;
    b. Major Injury;
    c. Fatality;
    d. Equivalent fatalities;
3. Safety loss to Neighbour:
    a. Minor Injury;
    b. Major Injury;
    c. Fatality;
    d. Equivalent fatalities;
4. Commercial loss (damage to property);
5. Delays to service;
6. Environmental damage.

The tool is based on the use of accident data from European railways to calculate estimated cost of safety, commercial and environmental losses.

The tool provides separate cost calculations for each of the above categories and includes the treatment of uncertainty and modular structure providing a flexible and expandable environment. The tool works in two modes:

*Site-specific:* allows the user to fully specify the region in which the incident occurs, including other input parameters such as types of train involved, time of day etc.

*Network:* the user is required to input only the type of incident and then the losses are calculated for averaged input parameters for a specified network (e.g. nationwide).

As result of that the risk prediction from the model is expressed in 15 categories corresponding to 15 loss categories as detailed above.

In order to support risk profiling, integration of individual core hazard models into one, overall safety risk model and options analysis, the existing technique cause consequence modelling was developed further as detailed in related sections below.

### 4.2.6 Model Data

Because the remit of the project was pan-European, (i.e. inclusive of all European railways) the model had to take into account all the different system configurations across Europe.

Hence, the only pragmatic solution to the problem was to develop a virtual railway, depicted by the averaged worst case parameters from across 5 European high speed railway lines.

Parameters identified as significant to our modelling were normalised to produce the European Reference Operational Environment (Lucic, 2005d). Following is a sample of used parameters:

1. Number of level crossings;
2. Number of stations;
3. Number of signals;
4. Average train speed;
5. Number of trains;
6. Number of passengers;
7. Average train weight;
8. Number of trackside workers.

These parameters were used to calculate failure rates for the model's elements, base events and barriers. For example, a failure rate of a subsystem is a function of the number of trains passed and their average speed and weight, or the probability of worker protection procedures being successful is a function of the average train speed, number of trains passing and number of workers on the line.

Collated together the information about parameters forms a Parametric Data Set (PAD).

Following from that, a data set containing all the parameters used to calculate the base events and barrier values were defined.

The parameter values from the data set were "fed" into the model, the calculations were run and "Crisp value" absolute risk for each consequence was calculated.

### 4.2.7 Integration of Individual Models into a Holistic System Safety Performance Model – Development

Initially, individual Core Hazard models were developed. However, the reality does not consist of a number of standalone hazards. Clusters of Hazards, referred to as Core Hazards interact with each other defining all the properties of the "Risk Space".

Therefore, it was necessary to identify all the dependencies between the Core Hazards. Some of the models where in fact, feeding into other models, defining the failure rates of higher level base events, or barriers as represented in Figure 4-3 below. Once all of the dependences were identified, the individual core hazard models were integrated into a "system safety performance model" (Lucic, 2005d).

In addition to the integration into a "system safety performance model", it was concluded that grouping all the similar consequences together into a so called 'virtual consequence' would be very useful; if the virtual consequences correspond to accident categories collated by industry or regulatory bodies since these can be used to calibrate the model against the historical data.

A simple criterion for grouping of consequences was developed specifically for that purpose:

| Total System Consequence | Level 4 virtual consequence | Level 3 virtual consequence | Level 2 virtual consequence | Level 1 virtual consequence | Real consequences |
|---|---|---|---|---|---|
| | Safety Consequences | Train Accidents | Movement | Collisions | |
| | | | | Derailments | |
| | | | | Level Crossing Accidents | |
| | | | Non-movement | In Stations | |
| | | | | Other premises | |
| | | Other (explosions, fires, structure failures, etc.) | Predominantly Passenger Related | NA | |
| | | | Predominantly Neighbour Related | NA | Consequences emerging from core hazards |
| Σ of all risks | | | Predominantly Worker Related | NA | |
| | Commercial / Environmental Consequences | Train Accidents | Delays | NA | |
| | | | Infrastructure Damage | NA | |
| | | | Environmental Damage | NA | |
| | | Other | Delays | NA | |
| | | | Infrastructure Damage | NA | |
| | | | Environmental Damage | NA | |

**Table 4-1: Consequence Groupings**

Consequences emerging directly from the core hazard models were grouped simply by summing all the same risk categories together. For example, predictions for two consequences are:

1. For consequence “A”: 2 fatalities per annum, 5 major injuries per annum and 34 minor injuries per annum;
2. For consequence “B”: 4 fatalities per annum, 7 major injuries per annum, 55 minor injuries per annum and £2,000,000 of commercial damage per annum.

An integrated consequence “C=A+B” is: 6 fatalities per annum, 12 major injuries per annum, 89 minor injuries per annum and £2,000,000 of commercial damage per annum.

The Total System Risk prediction is simply a sum of all the consequences together.

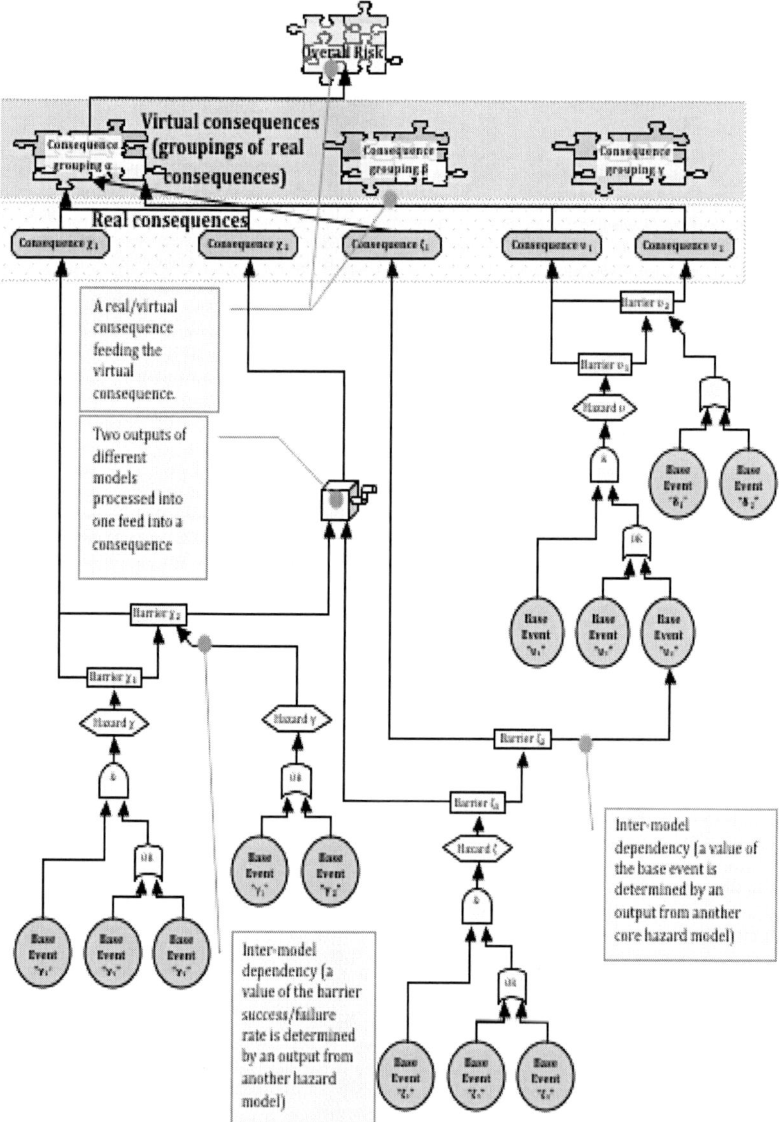


Figure 4-3: Core hazards models hierarchy

Four new functions and types of linking were identified during the integration process:

1. An output of one core hazard model determining the frequency of a Base Event, the constituent of another core hazard model structure. There are two possible ways that this link may happen:
   a. A link between two models is established/identified to pass the frequency of the consequence originating from one model to the base event of another model (the information passed is the calculated frequency of the consequence). This frequency becomes the frequency of the base event occurrence;
   b. A link between two models is established/identified to pass the frequency of the critical event originating from one model to the base event of another model (the information passed is the calculated frequency of the critical event). This frequency becomes the frequency of the base event occurrence;
2. A fault tree determining the probability of the success of a barrier, within one model. Sometimes a barrier is defined by a combination of events. In that case, a fault tree is required to describe and define the performance of the barrier;
3. A real/virtual consequence feeding the virtual consequence. If the linking between the models was established by the linkage of real consequences into virtual consequences, the information passed from the real consequence should be calculated risk for each loss category;
4. Intermediate processing. Regardless of the type of the link, information carried by the link could be processed in several different ways:
   a. Logical processing in terms of “AND” and “OR” logical operations performed on combined links;
   b. Splitting of links into several outputs with distributed weighting;
   c. Transform function performed on combined links producing single or multiple-weighted outputs;
   d. Simple addition of multiple links of the same type into a single link.

Following from the above identified novel functions, two different modelling environments were identified as well:

1. Individual models worksheets: Within this environment, individual models were produced. Basic characteristics of this environment are the same as the characteristics of the previous cause-consequence modelling environment, with the following exceptions:

   a. New inter-modal IN and OUT symbols were introduced, in support of integration and linking of outputs from one model as inputs into base events or barriers of other models;
   b. Facilities to support three dimensional classification of base events, as detailed later in section 4.2.10 Derivation of Requirements;
   c. An enhanced apportionment engine was incorporated into the model as detailed later in section 4.2.10 Derivation of Requirements;
2. Model Integration Environment: This is a completely new working/modelling environment developed within the ISAE tool.

New objects "operating" within this environment are first listed below and described in detail in APPENDIX B:

1. Critical Event Model super object (real consequences and base events of the model) to enable graphical linking of models into an integrated super-model;
2. Transfer Function intermittent operator;
3. Splitter intermittent operator;
4. Logical Gates (AND, OR) intermittent operator;
5. Super Connectors intermittent operator;
6. Virtual Consequences;
7. Links from real/virtual consequence to virtual consequence;
8. Links from real consequence to base event;
9. Links from the critical event to the base event.

These additional objects are necessary and sufficient to support the inter-modal integration process.

The inter-modal, model integration environment is illustrated by Figure 4-4 below. Following the integration of the Core Hazard Models, it was possible to calculate the contributions of individual model elements to the total System Risk.

Overall the integrated model consists of 40 individual models, corresponding to 19 core hazards (some core hazards could not be analysed by a single model, and were broken into smaller, more manageable pieces referred to as sub-core hazards). In total the integrated model consist of 583 base events, 240 states, 499 barriers and 365 consequences.

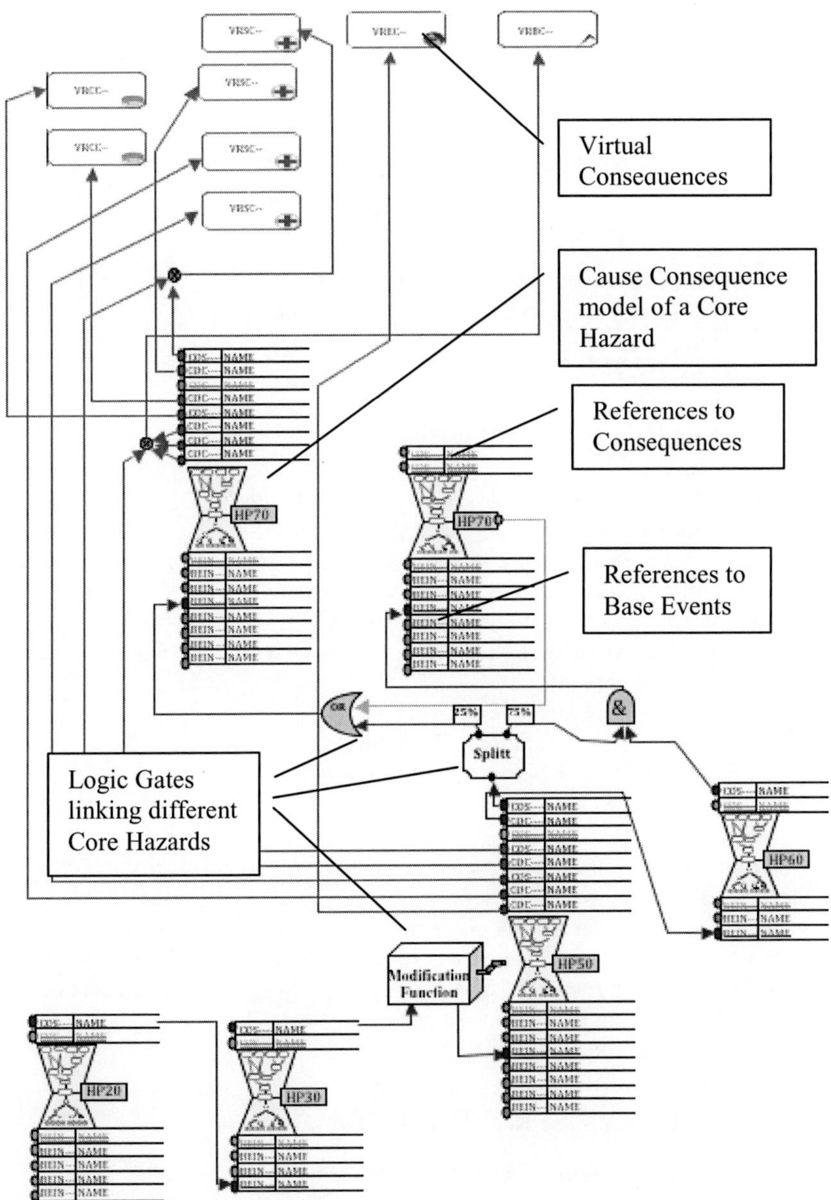


Figure 4-4: Model integration environment

### 4.2.8 Output Generation and Normalisation

The following is a sample of the forecasts:

| Consequences | | |
|---|---|---|
| Ref | Description | Annual Frequency |
| 1.1 | Collisions | 2.6 |
| 1.2 | Derailments | 1.1 |
| 1.3 | Other Movement Accidents | 0.2 |
| 1.4 | Commercial Only Accidents | 1600 |
| 1.5 | Null (no consequence) | 2900 |
| 1.6 | Level Crossing Accidents | 0.00088 |

**Table 4-2: Sample of ERTMS study result data**

Thus, the predictions are 2.6 collisions per annum, 1.1 derailment per annum, other movement accidents 0.2 per annum (or twice each 10 years), 1600 accidents resulting in commercial risk only each year, 2900 times nothing will happen though the hazard will exist (or in other words the accident will be averted) and 0.00088 level crossing accidents per annum.

Safety targets as stated by European Commission for Railways are (per year per exposed person):

1. Average population safety target –> 1.0E-05;
2. Passenger risk of fatality -> 3.3E-06;
3. Workers risk of fatality -> 3.3E-06;
4. Neighbours risk of fatality -> 3.3E-06.

Comparing the first forecasts against the above targets clearly indicates that the original model predictions for the safety performance of the system were unsatisfactory. A comparison of virtual consequences against railway accident reports was used to calibrate the models against the historical data. The review of data and logic resulted in several corrections to the logic and data of the model, until a particular version was accepted as representative (Lucic, 2005d).

### 4.2.9 Risk Profiling

The first step towards risk profiling was to define the parameters not as crisp values, but as statistical distributions, with minimum and maximum values and a distribution. Consequently calculated risks are articulated as statistical distribution. This type of risk profiling is referred to as "One Dimensional Risk Profiling".

By defining the parameters as functions of time and space, the risk calculations become three-dimensional, distribution functions of time and space. This type of risk profiling is referred to as "Three Dimensional Risk Profiling".

### 4.2.10 Derivation of Requirements

***Apportionment and Importance***

Although risk profile is extremely useful, being able to identify the contribution of model elements, or in other words, the contribution of equipment and procedural failures and human errors to the total risk is even more desirable. This process is referred to as "apportionment".

The following problem was encountered at the onset of the apportionment study.

Base events and barriers actually represent failures of equipment and procedures, which in different scenarios cause different problems; thus, what is required is to ascertain the absolute level of reliability of equipment and processes that must be achieved in order to produce a safe system. However, the integrated model is very large and complex, with many instances of the same equipment, or process failure causing different consequences.

For that reason it was necessary to find a way to distinguish the model elements that represent the same type of the failure.

As a solution to this problem, a three tier classification system was introduced, differentiating between:

1. A function provided by the system or constituent (primary classification), for example Train Speed Control;
2. A function provider (secondary classification), for example Automatic Train Protection System on board of the train;
3. A function environment (tertiary classification), for example failures arising During Protected Movement.

Therefore, part of the description of each model element (Base Event and Barrier) is its three parameter classification vector {x,y,z}.

For each classification group, primary, secondary and tertiary, a substantial number of classifications have been identified and the data on classifications is kept as a separate table depicting all the classifications and classification coding (abbreviations). Combining tables of classification coding with data sets containing parameter values (or functions) results in a data set referred to as a General Parametric Data Set (GPAD). The GPAD contains all the parametric information and classification descriptions feeding into the model.

Most of the configuration management of the model was concentrated on the GPAD. This is because once the model's logical structure was developed and confirmed as a correct one, this part of the model becomes almost static and the changes introduced to the model are mostly related to changes of parameter values and classifications.

The GPAD must be carefully configuration managed since a small change in one of the parameters may have large impact on the total risk prediction calculated by the model, and it is also necessary to know to what data set the calculations relate.

By simply changing the GPAD data, it is possible, using the same logical structure, to simulate different system configurations.

This is a great advantage, since until the system solution becomes stable, potentially many different system configurations are possible and representing each one separately through a direct change in the model would be extremely time consuming.

By adding together the contributions of all model elements with the same classification code, as well as the overall contribution of individual equipment, groups of people or procedures to the total system risk was calculated using new algorithms developed in support of this modelling technique.

Analysis was performed on 4 different levels for each risk category:

1. Apportionment of a virtual consequence to lower layer virtual consequences feeding it;
2. Apportionment of a virtual consequence to "source" consequences;
3. Apportionment of a virtual consequence to Core Hazards (Critical Events);
4. Importance of a virtual or a "source" consequence relative to Base Events and Barriers.

The author developed algorithms for all 4 calculations as outlined in the APPENDIX A.

### *Requirements*

Estimating risks emerging from a system is useful, but what is required is a target rate for each core hazard/sub core hazard or Tolerable Hazard Rates. For each of the core hazards, an annual rate that would satisfy the

Safety Targets, as specified by European commission for transport, has been calculated.

The hazard rate is expressed as an annual frequency of occurrence. Following on from that, the results of the modelling were used to establish the target Safety Integrity Levels (SIL) for Control Command Constituents (subsystems of the train control system). SILs were allocated to subsystems using calculated minimum and maximum annual failure rates and probability values (Lucic, 2005f).

### 4.2.11 Optioneering application

The most beneficial use of the modelling is in support of options and impact analysis. Using the model it is possible to identify the most important contributors to the risk, and to concentrate efforts to improve the performance of the system on these critical model elements (Lucic, 2005f).

An important part of options analysis is apportionment to the model elements. As already mentioned earlier, it is possible to distinguish three apportionment categories:

1. Apportionment to failures of functions within the system;
2. Apportionment to failures of function providers and
3. Apportionment to failures of function providers operating in specific mode.

For example, apportionment of total equivalent fatalities to equipment (control command constituents) and subsequently to functions provided by each control command constituent (the model indicates that 17% of the risk comes from the subsystem “L” (vital computer), where of total failures of the vital computer 35% happen during data entry, 19% during start up and test, 18% during train trip, 14% occur when protection for passing trains is activated and 16% occur when external alarms are activated).

There are two ways to utilise the model for optioneering and impact analysis. Firstly it is possible to change the parameters, values and shapes of their distribution, in order to see the impact of these changes on the model’s predictions. Secondly, a new base event or a barrier (a new model element) can be implanted and the impact of the change assessed.

In both cases of optioneering, it is the impact of the change, from original risk profile to new risk profile, that is being analysed. Using these results it is possible to justify either additional investment to implement new mitigation measures, or improve the performance of existing equipment, or reasonably reject implementation of the suggested mitigation measure using as an argument a line of reasoning that investment is grossly disproportionate to the benefits the new mitigation measure could create if implemented.

## 4.3 Second Project: Axle Counter modelling

### 4.3.1 Purpose and Scope of the Project

This section describes a risk modelling process as applied by the author to a real life project, from the initial stages of elicitation to final stages of testing and application.

Axle counter application risk models were developed to support the derivation of Safety Cases for individual axle counter installation projects. In accordance with the UK safety case regime, each project will have to prove that the technological and procedural solution accepted for the project is the As Low As Reasonably Practicable (ALARP) solution.

To achieve this, each individual project will have to demonstrate that a number of different options have been considered, and that the most suitable solution has been selected. In addition to this, the project will have to prove that the safety risks introduced to the railway system by the introduction of the new technology are acceptably low. In this section, the focus will be on the model that supported the analysis of the Reset processes, the Failures During Reset Model. The aim of this section is to:

1. Depict the development of Axle Counter Application Models;
2. Describe the process followed during the development of the models;
3. Outline the process for future use of models.

### 4.3.2 Axle Counter Train Detection Concept

The overall behaviour of an axle counter is represented by the state-space diagram in Figure 4-5. Essentially, an axle counter detects the passage of train wheels at the entrances and exits to a section of track, and increments or decrements a count accordingly.

If the count is more than zero, then the track section is set to occupied, and if less than zero the section is set to undefined. If the count is zero, and if internal safety checks, which will depend on the specific design of the axle counter in question, are satisfied, then the track section will be indicated as clear.

Thus, when the first wheel of a train entering the section passes over a detection point, the section will be indicated as occupied. As each subsequent wheel passes over the detection devices in the same direction as the first (usually called "counting heads"), the internal count will be incremented. As the train leaves the section, the count will be decremented for each wheel that passes over the counting heads.

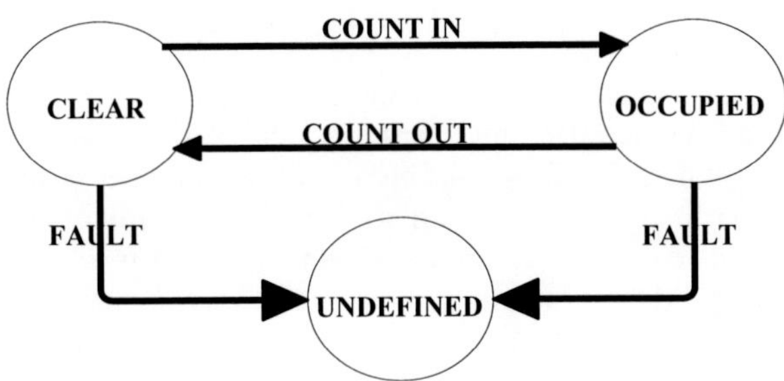


Figure 4-5: State-space diagram

When the last wheel leaves, the count will return to zero, and provided the internal safety checks have been satisfied, the section will once again be indicated as clear.

An important safety check, which is common in principle to all types of axle counter systems, concerns the detection of any event which could have interrupted the ability to detect and count wheels at any entrance or exit point. Such events might include interruption of the power supply, loss of communication with a detection point, or temporary removal of detection devices.

If such an event occurs, then it is possible that a train may have entered or left the section undetected, and it is essential that the axle counter should default to a non-clear state until it has been brought into correspondence with the actual state of the railway.

The default state, which is generally, designated the "disturbed" or "undefined" state of the axle counter system, will typically be treated as equivalent to an occupied state by the rest of the signalling system.

Recovery from the disturbed state can normally be achieved only by the operation of an external reset control, although it may be possible with certain designs of axle counter for the recovery to be achieved automatically following the successful completion of a complete count in – count out cycle as a sweep train occupies and clears the section.

The latter arrangement would be possible only for certain, very specific types of disturbance. The external reset facility resets the axle counter system to a clear state on all types of axle counter systems considered to date.

### 4.3.3 The Modelling Technique and the Tool

The technique used for the development of the risk models is known as Advanced Cause-Consequence Analysis and is described in a previous section. The tool used for development of these models is Integrated Safety Assurance Environment 8 (ISAE8). The Integrated Safety Assurance Environment assists development of Cause- Consequence risk models by organising the input, ensuring that all details captured are recorded, and allows the analysis of models using graphical worksheets.

Once the models are complete, ISAE8 can then generate a mathematical model and will allow different scenarios to be run to assess the influence of different factors on the model.

Whilst ISAE8 is a good analysis tool, it must not be the basis on which decisions about life and death situations should be based. However it does provide an insight into the model, and help to direct the focus of further studies and analyses required.

The ISAE8 tool is split into three distinct areas:

1. Pre-Processing;
2. Core Functions;
3. Post-Processing.

*Pre-Processing:* includes all of the functions required to set up the project management information and the initial data collection. The following are constituents of the Pre-Processing module of ISAE8:

*The Hazard Log,* should be used initially to record information about the Hazards.

*The System Parameters*. The tool supports parametric modelling (numerical values of model elements can be expressed as fixed numbers, parametric functions or statistical functions. This module of the tool allows for information about parameters to be defined. Sets of parameters can be organised in different Parametric Data Sets, corresponding to different scenarios to be modelled.

*The Classifications*. As the risk analyst construct the models, additional data is captured at significant points in the design in the form of classification codes.

When the model is complete the model can be calculated based on the apportionment of risk based on these classification codes. So when preparing ISAE8 for use in a new project, it is important to enter the Classification Codes early in the modelling process.

*The management* side involves setting up and maintaining information about:

1. Organisations involved in the project;
2. The hazards within the project;
3. The project team;
4. Project documents;
5. Project tasks.

*Core Functions:* includes the main Causal and Consequence modelling element of ISAE8, and allows the user to produce a risk model using various building blocks (Base Events, Logical Gates, Critical Events-Hazards, Barriers and Consequences), which can be drawn onto one or more Worksheets to model the Hazard.

Once the risk scenario has been drawn, it is validated; an Excel compatible Algorithm File (ALF) can be generated for subsequent simulation. It is also possible to define formulae for the losses pertaining to each Consequence or Accident. The Core Functions module also supports integration of models of individual hazard scenarios into one integrated, holistic model.

*Post-Processing:* allows further evaluation of the risk model by generating "What-if" scenarios. It also includes a reporting facility and the apportionment function to calculate the risks apportioned to selected elements.

Accurate and reliable train detection is an integral part of a safe railway signalling system. The principal method of train detection, in the UK, has been by the use of track circuits.

The widespread introduction on the UK Railways, of axle counters for train detection, where track circuits would otherwise have been used, is a strategic decision made by the Network Rail (UK Railways custodian).

Axle counters have only been used in specific applications in the UK, generally in small quantities. The large scale, widespread application of axle counters is new to the UK railway industry.

The Axle Counter Concept Safety Case (ATKINS, 2002) was developed to demonstrate that axle counter systems can be implemented safely on Network Rail controlled infrastructure where track circuits would otherwise have been used.

The Axle Counter Concept Safety Case does not attempt to show that risks associated with axle counters are ALARP in an absolute sense. It does show that the risks associated with axle counters can be made as low, or lower than, those associated with track circuits.

It provides a framework so that individual application safety cases can demonstrate that the type of axle counter system that is being commissioned on a particular project reduces the risks to ALARP (relative to provision or retention of track circuits) in a consistent and reproducible manner.

Application specific issues cannot be addressed at the concept level. To demonstrate ALARP, it is necessary to consider the specific application in conjunction with the Concept Safety Case, as the latter cannot deal with reasonable practicality in all situations and combinations of applications.

What is reasonably practicable is dependent on the specific technology of the axle counter, and of the signalling system to which it forms a constituent part, and the specific circumstances of the application.

Thus, individual Project and Application Safety Cases demonstrate that they have met the safety targets, the integrity requirements and the generic safety requirements provided within the Concept Safety Case, and that they have mitigated the specific project and application hazards. The Project Safety Case thus demonstrates that axle counters can be used safely as the sole train detection system where track circuits would otherwise have been used on Network Rail controlled infrastructure for that specific application but with the bounds dictated by the Concept Safety Case (ATKINS, 2002).

There are many factors to consider on UK railways infrastructure, all of which vary considerably throughout the UK:

1. Traffic density;
2. Rolling stock;
3. Traction systems;
4. Procedures;
5. Clearances; and
6. Control systems.

Consequently the model developed to support individual axle counter projects, had to be generic, independent of any particular project solution and application, but capable of simulating the particular project solution and environment.

### 4.3.4 Analysis Approach

Scope of analysis and modelling included all interfaces between the signalling system and the axle counter system, taking into account all the processes related to axle counter operation including control, command and communication exchanges.

The capability to realistically represent any chosen application is achieved by complete separation of logical structures and underlying data.

By changing the dataset used to run model calculations it is possible to simulate different, project specific solutions.

The process adopted for the development of models consists of 7 stages:

1. Review of existing related information;
2. Definition of the system;
3. Identification of hazards;
4. Analysis of hazards in terms of grouping and structuring;
5. Cause-Consequence analysis - Development of logical structures;
6. Cause-Consequence analysis - Parameterisation;
7. Testing, reviewing and gathering of data for testing.

The high level overview of the signalling system was developed in support of the identification of hazards and options for the mitigation of hazards.

The signalling system overview that was used as a base for hazard identification and modelling, defines all potential interfaces between the signalling system and the axle counter system, taking into account all the processes related to axle counter operation including control, command and communication exchanges.

The following are the primary interfaces with the axle counter system:

1. The rails;
2. Train wheels/axles;
3. Communications. [Note: the communications system(s) may or may not be part of the Axle Counter System, but are considered to be part of the train detection system];
4. Human Machine Interfaces (HMI), that provide the Reset/Restoration requests/commands and an indication of the status, (occupied, clear or undefined) of the track section and diagnostics;
5. Interlocking / signal controls including level crossing controls;
6. The power supply.

The operational environment for axle counters has been developed, and is described by, the "Typical operational environment and Application Diagram for Axle Counters" (Lucic, 2004a). This description includes all the elements of an operational environment; structures, passengers, staff, constituents of the signalling system, environmental factors, railway neighbours and their infrastructure, constituents of telecommunication and electrification, systems, trains, and engineering machinery.

### *Identification and Analysis of hazards*

Extensive and systematic hazard identification has been carried out based on the description of the system and the operational environment (Lucic, 2003b). All of the hazards identified during the hazard identification study have been analysed, structured and mapped to the Core Hazards (higher level hazard groupings) originating from the Axle Counter Concept Safety Case, which aims to support the development of Cause-Consequence models for the project.

Two Core Hazards have been identified as being related to the process of resetting the Axle Counter section:

1. Failures during reset (excluding level crossings);
2. Failures during reset (on level crossings).

Within the two core hazards, clusters of Causalities and Defences were identified. These were further analysed during the modelling stage of the project.

### *Model logic development*

Risk models were developed aiming to support any individual axle counter project. Consequently the models' logical structures are generic, independent of any particular project solution or application (Lucic, 2004a). The capability to realistically represent any chosen application, and to vary any relevant parameters, is achieved through complete separation of the logical structures and their underlying data.

By changing the dataset used to run model's calculations, it is possible to simulate different, project specific solutions. Cause-Consequence models developed for this project can be separated into two discrete, but completely integrated parts:

1. Cause-Consequence trees (logical structures);
2. Parameterisation of logical structures.

The models have been developed with the aim of supporting the comparative safety analyses of different procedural and technological options for the implementation of axle counter technology on different projects.

The models are inclusive of all potential system solutions. The following are areas where different options that may affect the outcome are likely:

1. The nature of the reset/restore procedure used;
2. Type of technical controls involved in the process;

3. The nature of the signal aspect given to the driver of the "next train";
4. The requirement for the signaller to apply protection to the section to be reset should occur before the reset, but has an effect when the incorrect reset has occurred;
5. Availability of different types of communication means between the train driver, signaller, technician and other railway staff.

Fault Tree methodology has been used for causal analysis. The causal parts of both models are almost identical, the only difference being a model element calculating the proportion of hazards on the level crossing section.

The fault tree is structured in three main branches:

1. The elements resulting in a failure whilst attempting to reset a section with more than one section requiring reset;
2. The elements resulting in a failure whilst attempting to reset a section with only one section requiring reset;
3. The elements resulting in a completely unintended reset of a section.

The consequence parts of the two models are different. For the model of "Failures during reset (excluding level crossings)" the Consequence tree is structured in three main branches:

1. Occupying train moving on plain line;
2. Occupying train moving at points;
3. Occupying train not moving.

The structure of the "Failures during reset (on level crossing)" model is very simple. If the section controlling the level crossing has been incorrectly reset, then the outcome depends on the presence or absence of road traffic, and on the actions of the road user and/or the driver of the "occupying train".

The model covers all types of level crossings - the likelihood of a particular level-crossing being affected by the reset of the axle counter and the potential to warn the driver of an approaching train that the crossing is not protected. For each model element, detailed descriptions are provided within the modelling tool environment. The frequency of occurrence of the hazard is measured in "number of incorrect resets per year" or frequency of failures during reset per year (how many times an occupied section has been reset each year).

***Parameterisation***

The use of parameterisation allows generic risk modelling of the Axle Counter Application. Whilst Cause-Consequence trees were developed to include all potential operational and engineering solutions to the application of axle counters in any geographical area, ease of customisation of models to a geographical area, or to a project, is achieved through parameterisation (Lucic, 2004a).

Parametric functions are used to evaluate numerical values (frequencies of occurrence and probabilities) of all the model elements. Parametric functions replace the crisp values for frequencies and probabilities of occurrence/failures of base events and barriers.

Numerical values of parameters within the parametric expressions are defined within the data-set. This technology allows for an easy and auditable way of changing the base data for the model and recalculation of estimates for different variations of the source data.

Data-sets are defined as lists of parameters used to calculate numerical values of model elements.

Parameters depict numerical values arranged in a number of data-sets describing different parts of the railway. Therefore, when the model is to be calculated, it is necessary to preselect the data-set.

This approach allows for increased flexibility in use of the generic model's logical structure. It also removes many of the undocumented assumptions hidden within fixed values that could result in an imprecise application model.

Parametric functions and data sets are used to analyse effects of the timetable related variations in operational performance, analysed scenario evolution elapsed time and spatial variations.

Some of the parameters, such as average headway and average speed, are related to the timetable and topology of the analysed route.

These parameters, if taken across the large area of the network, and as the difference between the minimum and maximum values of the parameter increases, introduce significant margins of uncertainty.

If parameters such as average headway and average speed, averaged across large area, are used to support the modelling, the outputs will be of sufficient quality to permit coarse estimation of risk fluctuation caused by replacement of track circuits by axle counters.

It also allows for elapsed time to be taken into account as values of relevant model elements are calculated by the parametric formulas containing elapsed time parameters.

The project only considered model structure, data issues and any inter-dependencies between these insofar as was necessary to derive, review and test the parameterisation. The main deliverable was the model database that was populated with parametric expressions, rationales for these expressions, and parameter definitions. The overall parameterisation methodology is illustrated by Figure 4-6. This methodology is described in detail as follows:

I. *Initial Review of Models and Documentation:*
A review was carried out of the models and relevant documentation including the Axle Counter Concept Safety Case and the Axle Counter Template Application Safety Case.

II. *Criteria:*
The aim of this model development work was to allow the axle counter risk models to support the safety arguments in the axle counter template application safety case.
The documents reviewed above were used in the development of these criteria.

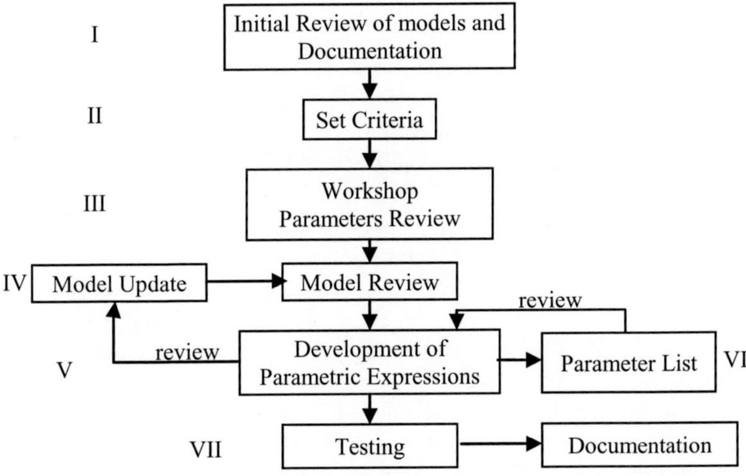


Figure 4-6: Parameterisation methodology

I. *Workshop:*
Having developed a set of criteria, a workshop was held to review the set of models.

II. *Model Update:*

The models underwent a series of iterative updates. These formed the input for the next phase of model review.

After each model update, the model was reviewed and parametric expressions developed.

A record of all the changes to the model is kept.

III. *Development of parametric expressions:*

Parametric expressions were developed based upon a review of the model structures. A rationale for each expression was drawn up and was given as part of the model element rationale.

IV. *Parameter List:*

A list of each of the parameters used in the formulae, together with their definitions and units, was produced.

This underwent a number of review stages to ensure that the parameter descriptions were at an appropriate level of detail. However it should be recognised that for some parameters, (particularly those that are simply a statement of a base event or barrier probability) a complete understanding of the scope or context of the parameters can only be provided by reference to the model structures.

V. *Testing:*

Testing of the expressions is described in more detail in the following paragraphs of this paper (Lucic, 2003c).

The parametric expressions were developed based upon the model review. A rationale for each expression was drawn up and given as part of the model element rationale. These were entered into the model database together with the formula rationale. For consistency, any parameter that represented a probability of success or failure was worded to represent failure.

Parameter units were normalised are as follows:

Time: hours or years; and
Length: kilometres

## 4.3.5 Tests

The complexity of the models developed during the modelling enhancement project dictated structured and comprehensive testing (Lucic, 2003c). The testing of the models encompassed the following model constituents:

1. Parametric Functions;
2. Logical Structures;

3. Model Predictions;
4. Data.

The required tests can be grouped into two categories:

1. Tests performed prior to application of the models to the particular project, encapsulating testing of Logical Structures and testing of Parametric Functions of the generic models;
2. Tests to be performed in preparation for and as part of the application of models to the particular project, encapsulating testing of Model Predictions and of Data.

In support of thorough and auditable testing of the models, a Test Plan was developed. Firstly, as part of the models' development process, all models were reviewed by domain experts at different stages of the project.

The final version of the models' logical structures was agreed between all involved as an accurate representation of reality. However, further reviews of the models should take place prior to their application on each specific project.

A second set of tests concentrated on Parametric Functions. The aim of these tests was to ensure the mathematical correctness of the Parametric Functions, and to eliminate logic errors in their definition. The data for these tests was supplied by Network Rail. This data was a coarse estimate of possible parameter and expected values for some model elements (Base Event, Barriers, Critical Events and Consequences).

In the absence of real life data, either for a specific application or for expected results, the aim of the testing was to show that the result calculated by the formulae was of the right order of magnitude.

The format of this data was as follows:

1. For each data item a logical Minimum value has been estimated;
2. For each data item a logical Maximum value has been estimated;
3. For each data item an average value has been estimated as a range.

Subsequently, each model was run using the above-described data sets. Any mathematical or logical inconsistencies that were discovered were resolved or explained.

The results of these calculations were compared against estimated expected values for model elements whose values are calculated from parametric functions (Base Events, Barriers, Critical Events and Consequences). All inconsistencies discovered were resolved or explained.

For each future application of the models to a real project, additional tests should be carried out to ensure the correctness of the models with regards to the particular application. These tests should be based on coarse project specific data elicited for testing purposes only. As a minimum, the data should be a coarse estimate of possible parameter values and expected values for some model elements (Base Event, Barriers, Critical Events and Consequences). The format of this data should be as described above.

Subsequently, each model should be run using the above described data sets. Any mathematical or logical inconsistencies discovered should be resolved or explained and documented.

Final tests should be carried out using real life data collected in support of a specific project.

In addition to this, a number of model elements calculated by parametric functions (States, Barriers, Critical Events, and Consequences) should be identified and where possible, values for these should be ascertained based on historical data or estimates. Subsequently, each model should be run, and the results of these calculations should be compared against the historical/estimated data described above. Any mathematical or logical inconsistencies discovered should be resolved or explained, and the process should be documented.

When applying the models to a particular project, whenever possible, the data for the modelling should be obtained from multiple sources and the best quality data should be used for the analysis.

However, some of the parameters needed for analysis are not readily available and will have to be estimated. In order to assure consistency of certainty boundaries for estimated parameters, it will be necessary to specify a common approach to estimation of parameters of similar nature, for example, all parameters related to human performance should be estimated using the same carefully specified process.

### 4.3.6 Application

Following the completion of all the above described tests, the gathering of data for a specific project was carried out. The first use of the models has been for the analysis of different options for the reset of an Axle Counter controlled Signalling Section after planned maintenance (Lucic, 2005a). Simulation of two options has shown significant differences in safety risk.

The first option involves the removal of the technical protections that were applied to the signalled section that is reset after maintenance. This option assumes that the first train coming along after the completion of the maintenance, and reset of the section, will enter the section without any technical caution. The following model elements have been affected by the first option:

1. Protection not properly applied before reset;
2. Route set for next train through affected section;
3. Driver of next train not cautioned;
4. Driver shown a special aspect;
5. Driver shown a red restricted by interlocking aspect;
6. Driver shown yellow restricted by interlocking aspect;
7. Driver pass auto signal at red but not cautiously;
8. No examination of the line with next train through (after maintenance).

The second option is the introduction of an unconditional reset facility to the system. Such a facility provides a means for a reset of the axle counter section, without any hardware or software self checking measures.

Only one model element, "unconditional reset facility used" is affected by this option. In both cases, procedural protections will be introduced to ensure that the section is clear (not occupied by trains, engineering vehicles, tools or other large objects) prior to reset and removal of the protections protecting the section.

Models have been run for three different sets of data:

1. Baseline data set containing data gathered during a number of structured workshops for a specific project;
2. Data set based on the baseline dataset reflecting implementation of the first option described above;
3. Data set based on the baseline dataset reflecting implementation of both, the first and the second option described above.

Comparison of the relative change of hazard and consequence frequencies related to changes being assessed, shows an increase in safety risk of two orders of magnitude.

In order to support the options analysis further, sensitivity analysis on the models should be performed in order to identify the model elements which contribute most to the risk. These are the model elements upon which to focus the analysis with the aim of improving the safety performance of the system.

These models have been developed to the point of test application. Prior to use of models for any project, testing of the models themselves should be carried out in accordance with the testing specification.

For each individual project, a test plan should be prepared. Sensitivity or importance analysis should be used to identify those model elements for which the overall risk is most sensitive, and particular emphasis should be placed upon the testing of these model elements. The parameters that are difficult to estimate accurately should be identified.

In order to assess the impact that accuracy of these parameters has on the overall risk, importance analysis should be undertaken. Particular emphasis should be placed on derivation of the critical parameters. Derivation of barrier probabilities should always be done using the model structures as an aid to analysis.

The use of graphical models increases the transparency of the modelling, and enables the user to see all of the factors which define the numerical values (failure rates/probabilities or frequencies/probabilities of events occurring) of model elements.

Parameters that define the barriers are contextual and conditional in nature, and it is therefore necessary to pay attention to the context of a barrier and related parameters during the data elicitation process.

## 4.4 Third Project: Complex Railway Project Safety Management – Manchester South Capacity Improvement Project

### 4.4.1 Purpose and Scope of the Project

The Manchester South Capacity Improvement Project (MSCIP) was undertaken as part of the West Coast Management Unit (WCMU) programme of works, with the purpose of providing additional capacity by upgrading the railway infrastructure in the South Manchester area. It was a multi-discipline Alliance Project covering track renewals and realignment, overhead line equipment renewals, power supply renewals and changes, and resignalling.

The project was realised in a number of stages:

1. The Stage A work was a 'Technology Demonstrator' for the application of the novel Italian signalling system, never before used on the Network Rail infrastructure. This stage was commissioned in April 2003;
2. The Stage A+ works built on Stage A with the implementation of the Train Protection and Warning System, and formed the next stage of the phased implementation of the signalling system. This stage was commissioned in December 2003;
3. The next phase was the Sandbach/Wilmslow (SHWW) Phase 1, with the purpose to close out Stage A/A+ issues and de-risk the SHWW Phase 2 stage, which followed;

4. When the SHWW Phase 2 stage was commissioned in summer 2006 the new signalling system was still not fully operational but the main functionalities were provided;
5. Final commissioning took place in spring 2007, with the complete set of functionalities provided as planned.

A Project Definition Document and an Outline Project Specification were developed, aiming to define the scope of each project phase.

### 4.4.2 Concept of Change Management

Each project has a duty of care to show that the railway as delivered is safe, and moreover to demonstrate that this has been done, must subject each change to the railway system to safety analysis. However, these changes vary enormously in significance, meaning therefore that safety analysis processes must be scalable. The initial remit for each stage of the project will require the Project to make certain changes to the railway.

As the stage proceeds, changes may be made to the remit as part of a formal change control process.

The total change made to the railway is a combination of the two, as the following diagram illustrates:

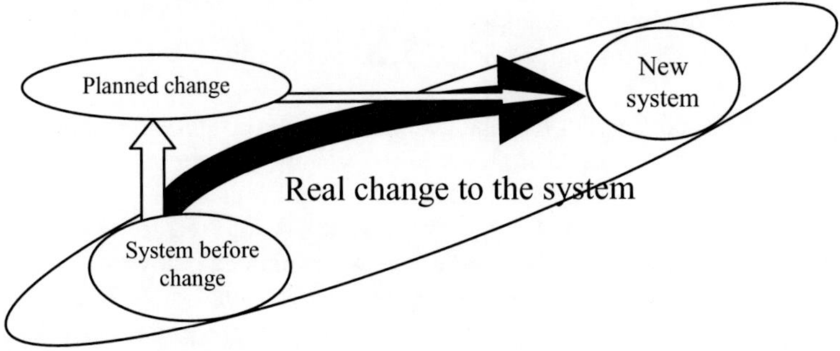


Figure 4-7: Concept of change

The safety analysis process for the project applies to:

1. All changes to be made to the railway which are defined in the initial remit for a project stage;
2. All changes subsequently made to that remit. These are typically described in Engineering Change Requests (ECRs). However the procedure also applies to non-engineering requests, such as

proposals to change maintenance regimes, which may be documented elsewhere or records within the DRACAS.

The analysis focused on identification of novel interfaces and their further safety analysis, where novel interfaces are defined as:

1. Interfaces between novel equipment;
2. Interfaces related to or controlled by novel procedures;
3. Interfaces between conventional equipment but operating within a novel environment;
4. Interfaces related to or controlled by conventional procedures equipment but operating within a novel environment.

All the interfaces are initially classified into one of four levels of safety significance. This process is referred to as Initial Change Safety Analysis (ICSA). The process categorises each change into four categories by safety significance, as follows (Lucic, 2003d):

Category 0: No hazards;
Category 1: All hazards controlled by existing standards or existing Hazard Log mitigations;
Category 2: Interface is novel but sufficiently simple and conventional that hazards can be analysed with reference to experience with existing equipment;
Category 3: Significant complexity and/or novelty.

Category 0 and Category 1 changes do not require safety analysis, and a generic process is defined for Category 2 changes. Category 3 changes have to be analysed according to their specific attributes. The process of analysis of Category 2 and Category 3 changes is referred to as Change Safety Analysis (CSA). The detailed process is explained below.

*Category 0 Changes:*_The Entry Criteria, Safety Analysis Process and recording requirements for the Category 0 changes are presented in the table below:

| | |
|---|---|
| Entry Criteria: | No credible condition or scenario can be identified in which the process of making the change or the changed part of the railway could contribute to an accident.<br>This is a strong criterion, which even very straightforward changes may fail, but will certainly apply to changes to the remit which do not affect the railway at all – for instance addition of a requirement to carry out a study. |

| | |
|---|---|
| Safety analysis: | None. |
| Records: | Reasons for categorisation. |

**Table 4-3: Category 0 Changes**

*Category 1 Changes:* The Entry Criteria, Safety Analysis Process and recording requirements for the Category 1 Changes are presented in the table below:

| | |
|---|---|
| Entry Criteria: | All hazards related to the process of making the change or the changed part of the railway are satisfactorily mitigated by authoritative good practice (such as Railway Group or Network Rail Company Standards) or by existing provisions within the System Hazard Log and the mitigation measures defined therein are put into practice. |
| Safety analysis: | None. |
| Records: | Hazards; mitigation measures (by reference to authoritative good practice or the Hazard Log); confirmation that these mitigation measures are being put into practice. |

**Table 4-4: Category 1 Changes**

*Category 2 Changes:* The Entry Criteria, Safety Analysis Process and recording requirements for the Category 2 Changes are presented in the table below:

| | |
|---|---|
| Entry Criteria: | The change is of small or moderate complexity and all hazards related to the process of making the change or the changed part of the railway are sufficiently similar to hazards in existing railway equipment that they can analysed with reference to experience with existing equipment. |
| Safety analysis: | A Change Safety Analysis Briefing Note should be prepared containing:<br>1. A description of the functionality of the change;<br>2. A description of the physical configuration of the changed part of the railway;<br>3. A description of the context of the changed part of the railway, identifying people, systems and |

| | |
|---|---|
| | equipment with which the changed part will interact.<br>A Working Group with sufficient knowledge and expertise should assess the safety implications of the change. The Working Group should analyse the change as follows:<br>1. Confirm that the change is correctly categorised and that the Group has sufficient competence to analyse it;<br>2. Review the System Hazard Log to identify previously defined hazards or hazard causes which relate to the change;<br>3. Review the intended functionality of the change to identify hazards or hazard causes which could arise from correct operation or plausible deviations;<br>4. Review the physical configuration of the change to identify hazards or hazard causes which could arise from failure of the constituents;<br>5. Review the context of the change to identify hazards or hazard causes which could arise from interactions between the change and its environment;<br>6. Identify from the System Hazard Log, the description of the change and knowledge of the existing railway, measures planned or in place to mitigate hazards or hazard causes identified;<br>7. Assess the adequacy of these mitigation measures and, if necessary, propose additional ones.<br>The process should be recorded in a Change Safety Analysis Report, and the System Hazard Log should be updated in line with the Change Safety Analysis Report. |
| Records: | Change Safety Analysis Briefing; Change Safety Analysis Report. |

**Table 4-5: Category 2 Changes**

*Category 3 Changes:* The Entry Criteria, outline of the Safety Analysis Process and recording requirements for the Category 3 Changes are presented in the table below:

| | |
|---|---|
| Entry Criteria: | Any change which does not meet the entry criteria for levels 0, 1 and 2. |
| Safety | The scope of hazard analysis activities related to Category |

| | |
|---|---|
| analysis: | 3 changes is to be defined for each change individually. The following information for each of the hazards should be collected prior to, during, or after the hazard analysis activity:<br>1. Reference to the hazard analysis undertaking;<br>2. Reference to the system/sub-system/interfaces being analysed;<br>3. Reference to a life-cycle stage or stages that the hazard is applicable to (e.g. a life-cycle should be precisely defined within the briefing documentation);<br>4. Hazard item identification number;<br>5. Hazard item description;<br>6. Description of causes of the hazard item;<br>7. Description of consequences of the hazard item;<br>8. Action related to further investigation of a hazard item (if required);<br>9. Additional comments;<br>10. Nominated authority/owner of the hazard (where it can be ascertained);<br>11. Reference to the analysis Briefing Note for Category 3 change including the following:<br>a. Session title;<br>b. Description of the system to be analysed (with clear definition of session boundaries/interfaces);<br>c. Description of all assumptions made;<br>d. Clear indication of the scope of the hazard analysis undertaking (including reference to the life-cycle and system states);<br>e. Explanation of the process and structure of the session including the clarification of the terminology to be used;<br>f. Competencies and expertise required for undertaking the session;<br>g. List of participants including an outline description of their expertise. |
| Records: | All hazards identified through hazard identification activities, as well as any hazards recorded/transferred from other sources (and where accepted by the Hazard Manager) must be recorded within the System Hazard Log. |

**Table 4-6: Category 3 Changes**

Cross references between the System Hazard Log entries related to novel interfaces and the system description were maintained.

A Project Definition Document (PDD) was reviewed, and ICSA was carried out for all of the changes planned for the project phase. Proactive safety analysis activities were then carried out for all the Category 2 and Category 3 changes. CSA was carried out on all changes in the PDD; these were split into two and analysed as separate activities. Changes/novelties analysed during the CSA were identified as differences between the railway system configuration corresponding to the previous phase of the project and the current one. However, the identified changes were then analysed against the standard railway system that was used as a reference.

All CSA (Lucic, 2003d) activities covered operations under normal conditions, degraded mode and emergency conditions (where appropriate), from a wider point of view, inclusive of transitions from and to the operational mode. While the boundary of the CSA was defined as the boundary of the project scope, the CSA activities also considered the operational interfaces with adjacent control areas, maintenance and emergency services.

As part of the activities, a review was carried out of the impact of changes to be introduced by a new phase of the project on the existing mitigations (identified and implemented during previous project phase), which aimed to confirm the validity of these mitigation measures.

Additional changes to those identified in the PDD were identified by Engineering Change Requests (ECR). These changes were reviewed, classified and analysed in accordance with the ICSA process. The results of all the analyses and related actions were kept in the ICSA register. All DRACAS records that were identified as having safety implications were referenced in the System Hazard Log. In cases were a safety related DRACAS entry did not correspond to any of the existing System Hazard Log entries, it was subjected to safety analysis.

### 4.4.3 System Scoping and Conceptualisation

The system was depicted in the "System Architecture" document. The "System Architecture" describes all significant subsystems and internal and external interfaces of the system. The system architecture is presented in three levels of detail.

Starting with the highest level, depicting interfaces with objects outside the project scope and outside of the railway itself; the second level provides a description of the main subsystems and their internal (within the scope of the project) relationships; and finally, providing the detail internal to the subsystem at the 3rd level of the description.

The system architecture was accompanied with an Objects & Interfaces database which described all objects and their interfaces. Detailed information about the changes to these interfaces throughout the project life (for each stage) was kept in this database. This information was then used to support ICSA and CSA sessions. An illustrative example of the "System Architecture" diagram is on Figure 4-8 below.

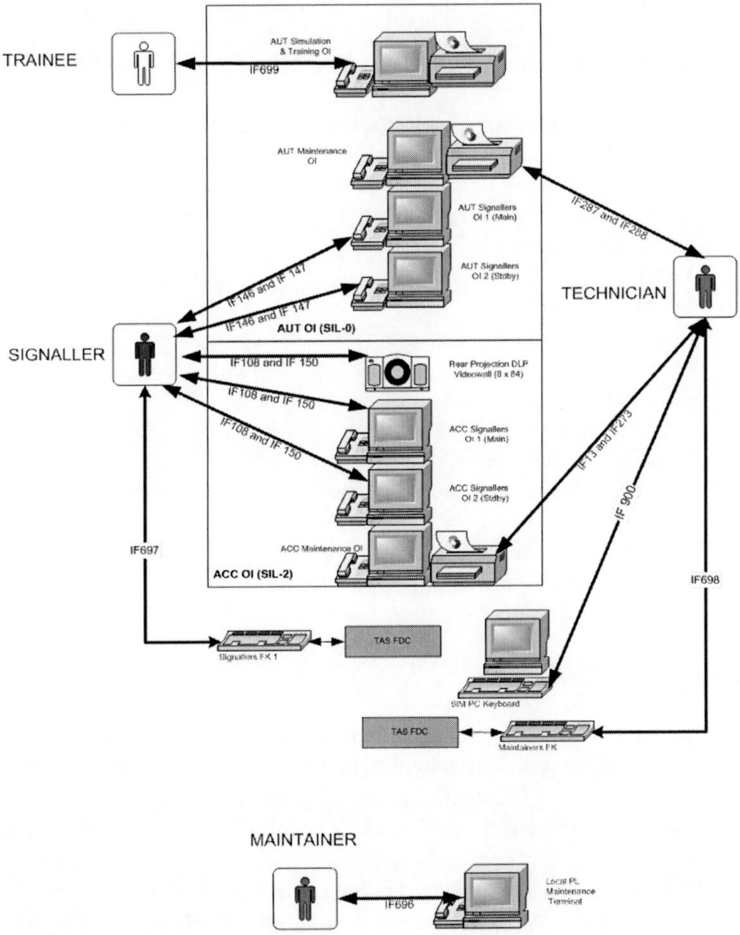

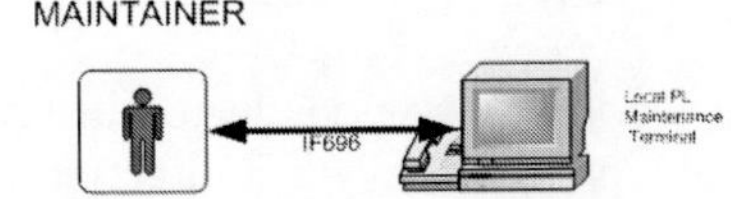



Figure 4-8: An illustrative example of the "System Architecture" diagram

### 4.4.4 Identification and Analysis of Hazards

The identification of Hazards was structured in the three parts (Lucic, 2004b):

1. Characterisation of Change and Identification of Safety Benefits;
2. Hazard Identification – Human Factors and Operational Issues;
3. Identification of Hazard Mitigations and Actions required which was undertaken in parallel with Hazard Identification.

#### *Characterisation of Change and Identification of Safety Benefits*

The purpose of this part of the study was to agree the scope, extent and purpose of the change in a form suitable for analysis at later stages, and to identify the safety benefits of the change.

This was achieved by considering the following questions:

1. What is the change?
2. What is the rationale behind the way in which the changes are applied?
3. What is the intended function of the new equipment?
4. What is the environment in which the new equipment will be installed?
5. To what does the new equipment interface?
6. What non-equipment changes, if any are being made at the same time? (For instance, changes to maintenance regime or operational rules).
7. Which aspects of the equipment and its application are novel and which have been proven in service?
8. What are the safety benefits?

#### *Identification of Hazards*

A hazard is any possible state or behaviour of the signalling system, or operator interaction with any inherent equipment within the system, which might contribute to an incident. Actions of passengers and members of the public were not considered by CSA as these interfaces were outside of the project scope.

A brainstorm workshop was conducted in order to identify hazards, during which the changed functions, interfaces and equipment were considered. It was considered that the topic was well understood and that a structured brainstorm would be an efficient method of analysis.

The standard guide words, listed in Table 4-7 below, were applied to each interface affected by the change as described above. This provided structure to the hazard identification process. The fact that a potential hazard was fully mitigated was not grounds for removing it. It was recorded in order to take full credit for the mitigation of the Hazard.

| Guideword | Meaning | Sample Interpretations |
|---|---|---|
| NO or NOT | The complete negation of the intention | No execution, missing information, not ready, delay, earlier or later |
| MORE/EARLY | Quantitative increase | Parameters & items - too many time & range - too high, too long (both singly and in combination) |
| LESS/LATE | Quantitative decrease | Parameters & items - too few time & range - too low, too short (both singly and in combination) |
| AS WELL AS | Qualitative increase | Parameters - extras, incorrect input, extra extension, different format actions - too many, too often, redundancy (in functions) |
| PART OF | Qualitative decrease | Incomplete input/operation, truncated, execute part of |
| REVERSE | The opposite of the intention | Opposite sign, feedback, inverse, negative parameters, go back |
| OTHER THAN / WRONG | Substitution of the intention | Intended input, intended operation, redirection of input, similar command, configuration control, wrong manual, wrong file, wrong value, unrequired act performed. |

**Table 4-7 : Hazard identification Standard Guide words**

### *Mitigation of Remaining Hazards*

The purpose of this part of the study was to identify any existing and planned, technical and procedural measures that mitigate the hazards remaining after the analysis performed above.

The following is an example of the outcome of the CSA workshop for an interface: IF 13 (Diagnostics): From ACC Subsystem to Technician.

| Characterisation of Change and Identification of Safety Benefits | |
|---|---|
| What is the change? (Confirm correctness of information in the Hazard Identification Study Briefing Document) | IF13.1. (P1): UK1 platform will have increased diagnostic. The maintainer gets a new alarms window which;<br>1. shows the alarms on the system as they happen, and displays a hierarchical version of the system which can be navigated<br>2. allows the maintainer to sort the alarms<br>3. allows the maintainer to suppress alarms from a specific part of the system, thus reducing the number of irrelevant alarms<br>4. notifies the maintainer when the alarm status is not available or reliable for a specific part of the system<br>5. notifies the maintainer when there is an alarm present in a specific part of the system<br>6. notifies the maintainer if the alarm is intermittent - i.e. comes and goes, thus reducing the number of irrelevant alarms displayed<br>7. suppresses alarms on equipment where a fault in the system makes the alarm status for that equipment unavailable or unreliable<br><br>(Alarms are expected to latch. It is to do with ACC only no On-Line Diagnostic Terminal (i.e. SIM PC)<br>(Action on xx: Confirm that alarms for UK1 (Phase 1) are latching (as defined in the alarms specification)<br><br>IF13.2. (P2): number of screens increased, screen configuration changed |

| Characterisation of Change and Identification of Safety Benefits | |
|---|---|
| What is the rationale behind the way in which the changes are applied? | DRACAS<br>Different management of alarms needed |
| What is the intended new function? | Improve alarms management |
| What is the environment in which the new equipment will be installed? | Same as before |
| What does the new equipment interface to? | Same as before |
| What non-equipment changes, are being made at the same time? (e.g, maintenance regime changes, operational rule) | Maintenance regime of monitored equipment<br>Possible change to operational rules should the equipment as implemented not function according to current Rule Book assumptions |
| Which aspects of the equipment and its application are novel and which have been proven in service? | Novel to the UK, used in Italy. |
| Safety benefits of change | For IF13.1. (P1): Improved maintenance gives better system availability<br>For IF13.2. (P2): Optimal visual presentation of information |

**Table 4-8: Characterisation of Change and Safety Benefits**

And finally an example of the records table for the above interface, from the workshop:

| Change Ref | Hazard | Cause | Mitigation | Action | Guide word |
|---|---|---|---|---|---|
| IF13.1 | No immediate hazard (Failure of diagnostics needs to be rectified as required) | Failure of diagnostics (no or wrong alarms) | | | No(info), part of |
| IF13.1 | Unable to distinguish critical alarms | Too many alarms | Alarms level & thresholds should be set at appropriate level | J Bloke- Ensure that this is correct | Too early (report failures before they happen) |
| IF13.1 | Not use the system properly, shift to unnecessary degraded mode of operation | Use of over restrictive controls | Design and testing of system to appropriate standards | | Wrong (info) |

**Table 4-9: An example record from the CSA workshop.**

### 4.4.5 Comparative Risk Assessment

All of the hazards were grouped in two categories:

1. The risks associated with the hazards that are understood and currently mitigated as standard practice; i.e. the risks are mitigated by following standards or other established authoritative good practice, and the project is actually following these standards or

other authoritative good practice. These hazards were not analysed further.

2. Novel hazards that originate from novel equipment, interfaces, or procedures, or from conventional equipment or procedures operating within the novel environment. These hazards were analysed further.

The risk associated with each novel hazard was categorised as follows:

BETTER: if the risk is assessed as significantly better than would be achieved with established conventional (like for like) means ( in case of railway signalling for example with existing Solid State Interlocking, conventional UK signal heads etc);

OR

WORSE: if the risk is assessed as significantly worse than would be achieved with conventional (like for like) means;

OR

COMPARABLE: if the risk is assessed as neither Better nor Worse than would be achieved with conventional (like for like) means;

*AND*

MINOR: if there are other hazards associated with the scheme that have a risk at least an order of magnitude greater;

OR

MAJOR: if there are no other hazards associated with the scheme that have a risk at least an order of magnitude greater.

The risk associated with a hazard is considered acceptable only if it is (MINOR and COMPARABLE) or (MINOR and BETTER) or (MAJOR and BETTER).

In general, the Hazard was declared as closed (managed to sufficiently safe level) when:

1. The risks associated with the hazards could be shown to be comparable with, or better than, what would be achieved with a conventional scheme and no further reasonably practicable mitigation could be identified; or

2. In the case of standard hazards or hazards originating from either the conventional systems already in use or standard procedures already applied on the railways, the closure of these hazards is justified through the application of well-established good practice, including Railway Group Standards and Network Rail Company Standards. These hazards were not further assessed.

Within the project timescales, options analysis concentrated on hazards classified as (MAJOR, COMPARABLE) which account for the majority of the risk. Each (MAJOR, COMPARABLE) hazard remaining at the end of the project phase was subjected to an options analysis in order to underpin an argument that there are no reasonably practicable tactical options available to reduce risk further.

A list of factors that act to drive risk up or down with respect of a conventional UK resignalling system was collected in support of a qualitative but rigorous argument that risk would not be reduced in the long-run by the strategic option of replacing the ACC with a conventional UK resignalling.

The ALARP argument rests in part on the conclusion that there are no tactical or reasonably practicable strategic options, which would reduce risk further. As part of the options analysis, a list of performance metrics that would allow the project, to make a quantitative assessment of the risk associated with (MAJOR, COMPARABLE) hazards at a later stage was identified.

A recommendation to Network Rail to collect performance metrics was made.

### 4.4.6 Management Process

This section describes how project hazards were identified, analysed and tracked to formal closure.

Figure 4-9, below describes the Hazard Identification, Analysis and Management Process.

Inputs to the process are:

1. A detailed description of the interfaces extracted from the system Breakdown Structure. All interfaces were subjected to ICSA process as described above. To support the process it is necessary to understand the nature of the interface, its performance limitations, operating environment, etc. All this information should be provided as part of interfaces description;

2. A detailed description of any change to the system that is made subsequent to changes defined in the initial scope of the system. This was done through Engineering Change Request process, where each change was recorded and subjected to an ICSA process;
3. DRACAS process records were subjected to ICSA process as well. A review of the DRACAS records served a dual purpose. Firstly, to assess the safety impact of any changes to be introduced to the system and secondly to confirm the correct functioning of the identified and implemented mitigation measures.

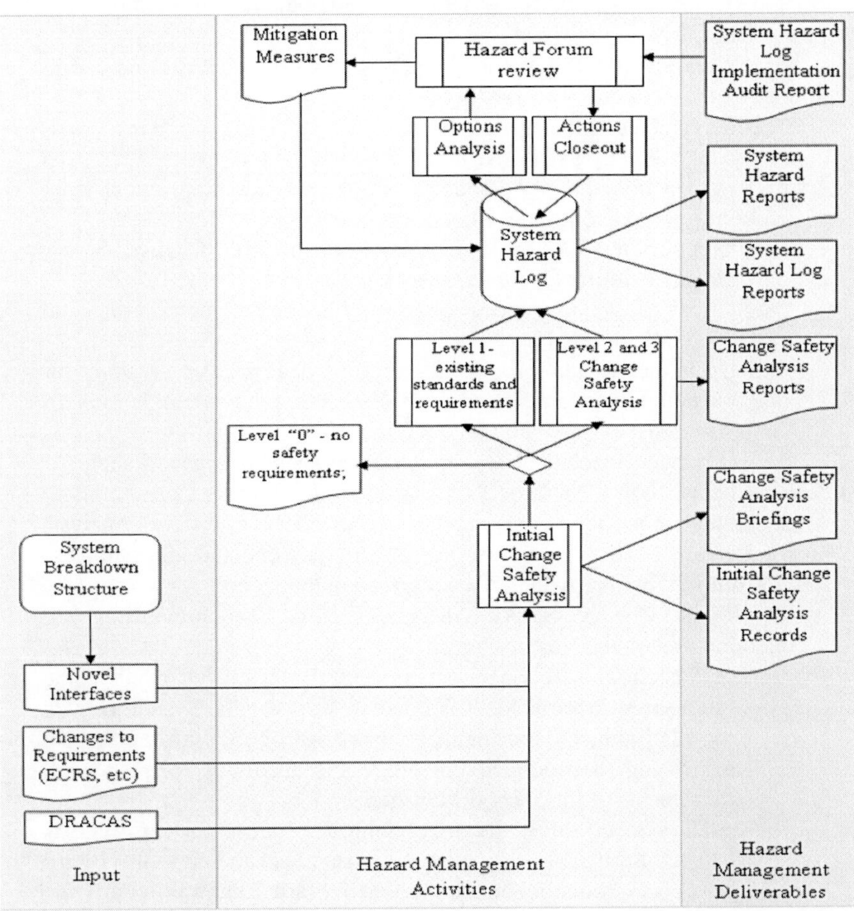


Figure 4-9: Management Process

Each change was subjected to ICSA, Category 2 and 3 changes were subjected to CSA. The results of the CSA were kept in the System Hazard Log, as well as references related to Category 1 interfaces. The System Hazard Log was used as a tool to support the management process.

Each of the identified mitigations has been allocated to a responsible person on the project for implementation. Proof of implementation was referenced/recorded in the System Hazard Log. These were subjected to review by a team of suitably selected experts, to confirm satisfactory implementation of the mitigation measure. Upon agreement from the Hazard Forum (Lucic, 2002), the action to implement the mitigation would be closed.

Each hazard was subjected to an Options Analysis process as described above.

The deliverables of the process were:

1. Initial Change Safety Analysis Records: ICSA of an ECR was recorded in a spreadsheet and these records were kept under strict configuration control;
2. Change Safety Analysis Briefing Document(s) Change Safety Analysis Briefing Documents were produced to:
   a. Provide practical details such as location of the meeting and agenda for the day;
   b. Summarise the process to be followed for the session; for example the process to be followed for the identification of hazards;
   c. Provide a technical description of the scope of the workshop.
3. Change Safety Analysis Records: The output of the Change Safety Analysis activities were recorded in Change Safety Analysis Records. These records accurately documented the process followed to clearly identify the agreed output.

   The Change Safety Analysis records were reviewed and approved by the Hazard Forum.
4. The System Hazard Log was the central repository for all of the system safety hazards identified for the project. The System Hazard Log was produced and maintained in accordance with the Project Hazard Log Management Procedure and according to the overall requirements of the Hazard Management Plan.

   All system safety hazards identified across the project, by whatever means, were tracked to closure via the System Hazard Log. A key function of the System Hazard Log was to provide traceability from each hazard to all evidence used to justify its elimination or mitigation.

5. System Hazard Reports were issued at key milestones throughout the project, reporting on the status of all hazards within the System Hazard Log. These reports include a summary of the number of hazards in the System Hazard Log for each hazard classification (e.g. open, closed, mitigated etc).
6. The System Hazard Log Report contained the following information:
   a. System Hazard Log Section, which describes the purpose of the System Hazard Log, its context in the history of the Project, the development of the System Hazard Log, its scope, use and underlying assumptions and constraints as well as a key to the fields in the Hazard Record Sheet and Hazard Summary List;
   b. Journal of changes to the System Hazard Log;
   c. Summary list of hazards and their status;
   d. Hazard Record Sheets contain the following information:
      i. Information on the hazard: its unique reference number, description, and status. Information was also provided on the hazard's worst-case consequences should it occur, and the assessed frequency, probability and consequences of occurrence.
      ii. A cross-reference was provided to the hazard's original source;
      iii. Each cause of the hazard was uniquely numbered (with numbering derived from the hazard's number) and described;
      iv. For each cause a listing of the actions to be taken and mitigations already in place was also provided;
      v. Each action was uniquely numbered (with numbering derived from the relevant cause of the hazard) and described.
      Each action was allocated a nominated person to resolve the action, together with information supporting the action taken and/or mitigation proposed.
   e. Directory of safety documents on the project.
   f. List of consequences identified for all the hazards in the System Hazard Log.
7. The System Hazard Log Implementation Audit Report presented findings of the Project Hazard Log Implementation Audit (Lucic, 2003d).

# 4.5 Fourth Project: High Level Railway System Study - POLAND

## 4.5.1 Purpose and Scope of the Project

'The Technical Assistance for Preparation of the Modernisation of Corridor II – Remaining Works' project was aimed at enabling the Government of the Republic of Poland to make strategic decisions and financial commitments regarding the Modernisation of the Corridor II railway link. The geographical scope of the project includes the E20 section of the railway, Warszawa – Rzepin (excluding the Poznań node), and the CE20 railway section, Łowicz – Skierniewice – Łuków.

The Technical Assistance project was realised in 5 stages:

Stage I: Feasibility Study - Cost Benefit analysis of three proposed options;
Stage II: Feasibility Study – Safety, Environment, Economic and Financial Analysis;
Stage III: Preliminary design;
Stage IV: Preparation of Application for EU funds;
Stage V: Elaboration of Tender documentation for contracts.

The author led the work at Stages I and II. Three potential options were suggested:

Option 0: Basic construction work required to provide a minimum set of requirements with maximum limitations, resulting in a conventional railway line achieving the ERTMS compatibility at operational level only;
Option 1: Modernisation and upgrading of the infrastructure to Technical Standards for Interoperability of Conventional Rail System and AGC and AGTC standards for international corridors (maximum speed of 160km/h for passenger trains and 120km/h for freight trains). Therefore this option will mean introduction of ERTMS Level 2 railway line;
Option 2: Extension of option 1 to enable maximum speed 200 km/h, assuming use of tilting passenger trains, resulting in introduction of ERTMS Level 2 railway line as well.

For each Project Option, a number of different System Solutions was suggested. As part of the activities within the first two stages of the project, a cost benefit analysis of the proposed three options and system solutions was carried out. In order to achieve this, it was necessary to identify relevant, high level, safety, RAM and environment protection requirements for each of the three proposed project options, and to identify potential the benefits that each proposed project option could bring. In terms of compliance with EU standards and directives with regards to safety, the project had to, broadly, follow guidance provided in EN50126 and to be compliant with Control Command and Signalling (CCS) Technical Specification for Interoperability (TSI) Mandatory Specifications as follows:

1. Index 27: UNISIG SUBSET-091 Safety Requirements for the Technical Interoperability of ETCS in Levels 1 & 2;
2. EN50126: Railway applications - The specification and demonstration of Reliability, Availability, Maintainability and Safety (RAMS). This standard is applicable to the whole of EU;
3. Other Specifications/Standards/Documents mandated by the TSI's;

   and Informative Specifications as follows:

   a. B43: 04E083 "Safety Requirements and Requirements to Safety Analysis for Interoperability for the Control-Command and Signalling Sub-System". Being a type "1", these specifications represent the current state of the work for the preparation of a mandatory specification still "reserved";
   b. B44: 04E084 "Justification Report for the Safety Requirements and Requirements to Safety Analysis for Interoperability for the Control-Command and Signalling Sub-System". Being a type "2", these specifications give additional information, justifying the requirements in mandatory specifications and providing help for their application.

Regarding RAM, there are documents produced in relation to ERTMS which are similar to the forms of safety that the projects have to comply with and refer to.

### 4.5.2 The Process for derivation of Safety & RAM requirements & Hazard Identification

Process-wise, the derivation of safety and RAM requirements was carried out in two steps, corresponding to the first two stages of the project, Stage I and Stage II:

Step 1: Identification of the high level safety, RAM and environment protection requirements for each of three proposed project options, as well as the benefits that each proposed project option, could bring to the system, for inclusion in the high level requirements baseline. At this stage of the project, a very high level set of qualitative system requirements/policy statements, cross referenced against related standards (listed above), was identified, aiming at the synchronisation of project requirements with local Polish and ERTMS standards and high level requirements. For this step the requirements were defined in relation to the overall system performance. The Polish railways do not have established safety targets for the railway network or railway lines, and therefore the project had to propose the approach to the derivation of the overall safety targets to Polish railway authorities (PKP) and the Safety Authorities. Since the E20/CE20 railway line is part of TEN lines, and the line is to be equipped with ERTMS Level 2 railway control system, it was logical to expect the safety performance to be at the same level as that of other ERTMS fitted or planned lines in EU.

Consequently, the project made use of work done by UNISIG (RAMS target set at $10^{-9}$ dangerous failures for on board and track side ERTMS/ETCS equipment). Following the logic that safety performance for a conventional signalling system must be same as that of ERTMS hence, the overall safety target for a conventional system should be $10^{-9}$, i.e. SIL4 (Safety Integrity Level 4).

Step 2: Derivation of more detailed requirements specification in support of the tender documentation entailing apportionment of the identified/agreed overall system RAM and safety requirements to the main system constituents. The RAM requirements were set at a different level of system decomposition to safety targets. At this stage, a more detailed set of safety requirements were derived. In order to do that, the project had to develop a high level functional system description inclusive of ERTMS and conventional system constituents, and to identify interfaces for which the safety and the RAM targets will be identified. Identification of the interfaces for which the project needs to define requirements, was dependant on the contracting/supply policy adopted by the project. Using that system description, the HAZID study was organised in order to identify high level hazards which were mapped or cross referenced to existing lists of CCS hazards as defined in B43: 04E083 and B44: 04E084. Other, non CCS parts of the railway system and related hazards were treated separately. This resulted in a complete list of railway system hazards, and forms an initial hazard log for the project. The outcome of Step 2 activities was:

1. A paper, for inclusion in tender documentation, stating:
    a. References to relevant standards and directives;
    b. System description used in support of the work;
    c. List of RAMS requirements;
    d. List of Assumptions, Dependences and Caveats;
    e. List of identified system hazards.
2. A hazard log:
    a. Hazard log in "paper" format for inclusion in tender documentation;
    b. Hazard log database for handover to railway authorities and contractors.

### 4.5.3 System Scoping and Conceptualisation

In support of the first step (as defined above) a high level of system conceptualisation was developed (Lucic, 2006). It is possible to define 8 classes of different railway system constituents:

1. Controlled infrastructure;
2. Maintenance, Tools, equipment, machinery;
3. Trains;
4. Stations;
5. Stabling areas;
6. People (Procedures);
7. Railway neighbourhood;
8. Environmental factors.

Each of these can be broken into more detailed subclasses, as follows:

| Class/Subclass | |
|---|---|
| 1. | Controlled Infrastructure |
| 1.1 | Civil structures |
| 1.2 | Infrastructure based Communication Systems |
| 1.3 | Electrification |
| 1.4 | Level Crossings |
| 1.5 | Permanent way |
| 1.6 | Plant |

| Class/Subclass | | |
|---|---|---|
| | 1.7 | Signalling (Conventional) |
| | 1.8 | Signalling (ETCS trackside) |
| | 1.9 | Infrastructure in general |
| 2. | Maintenance, Tools, Equipment and Machinery | |
| | 2.1 | Maintenance Regime |
| | 2.2 | Manual Tools |
| | 2.3 | Instrumentation |
| | 2.4 | Spares |
| | 2.5 | Power Tools |
| | 2.6 | Machinery(e.g. diggers) |
| 3. | Trains | |
| | 3.1 | Braking Systems |
| | 3.2 | Train Based Communication Systems |
| | 3.3 | Doors |
| | 3.4 | Running Gear |
| | 3.5 | Traction systems |
| | 3.6 | Train - Materials of Construction |
| | 3.7 | Train Control (e.g. AWS, TPWS) |
| | 3.8 | Trainborne ETCS |
| | 3.9 | Auxiliary Systems and Supply |
| | 3.10 | Freight |
| | 3.11 | Rail Based Machinery (e.g. ballast cleaners, tampers) |
| | 3.12 | Material/Fuel |
| | 3.13 | Other (i.e. technical or undetermined issues) |
| | 3.14 | Corporate responsibility |
| 4. | Stations | |
| | 4.1 | Escalators |
| | 4.2 | Lifts |
| | 4.3 | Non-public areas |
| | 4.4 | Platform areas |
| | 4.5 | Other station areas |

| Class/Subclass | | |
|---|---|---|
| | 4.6 | All station areas |
| 5. | Stabling Areas | |
| | 5.1 | Depots/workshops |
| | 5.2 | Sidings |
| 6. | People (Procedures) | |
| | 6.1 | Control room staff |
| | 6.2 | Emergency response staff |
| | 6.3 | Neighbours |
| | 6.4 | Passengers |
| | 6.5 | Station staff |
| | 6.6 | Infrastructure maintainers |
| | 6.7 | Train crew |
| | 6.8 | Train maintainers |
| | 6.9 | Operational Staff |
| | 6.10 | All Staff |
| | 6.11 | Corporate responsibility |
| | 6.12 | Axle Counter operation specific procedures |
| | 6.13 | Training |
| 7. | Railway Neighbourhood | |
| | 7.1 | Construction |
| | 7.2 | Hospitals |
| | 7.3 | Housing |
| | 7.4 | Industry/Factories |
| | 7.5 | Roads |
| 8. | Environmental Factors | |

**Table 4-10: Railway system constituents classes and subclasses**

In support of the Second step (as defined above) a more detailed, but still high level, system conceptualisation was done as illustrated in Figure 4-10 below.

A system boundary for the analysis was identified, and the relevant interfaces marked by different colours in order to denote different suppliers delivering different system constituents. All coloured interfaces are the boundary interfaces. Each identified boundary interface was reviewed in order to identify those interfaces (for each project option) that are not covered by existing standards, and for which it is necessary to identify the requirements and related potential hazard. The following table contains a sample of this information.

| Requirements & hazards to be identified/ Interface number | | Interface between | |
|---|---|---|---|
| | | Object | Object |
| Yes | I01 | Infrastructure Condition Monitoring | Infrastructure failure data |
| No | I02 | Maintainer workstation | Infrastructure Condition Monitoring |
| Yes | I03 | Trackside ERTMS equipment status | Infrastructure Condition Monitoring |
| No | I04 | Functionality of Level Crossings | Infrastructure Condition Monitoring |

**Table 4-11: An example of interfaces description**

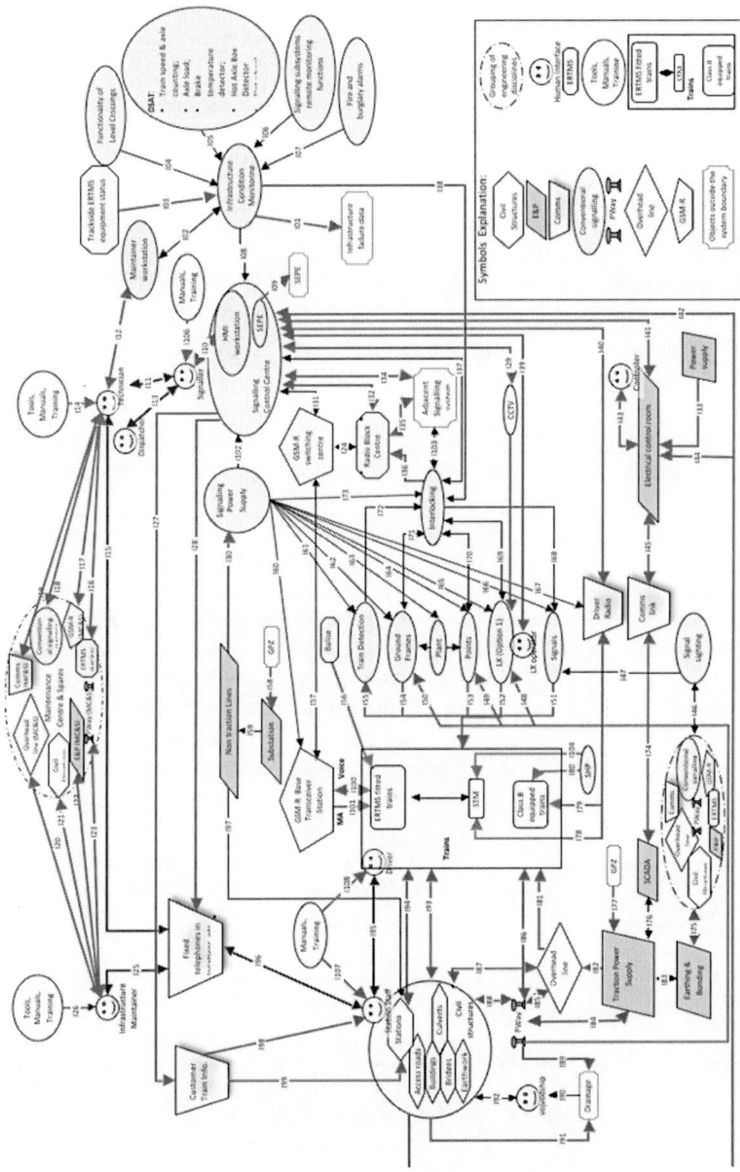

Figure 4-10: An illustrative example of a Railway System Model

### 4.5.4 Derivation of High Level Safety Requirements

For each option, all of the classes listed in Table 4-10 above were reviewed in order to identify the high level safety, RAM and environment protection requirements and benefits. The analysis was focused on identifying only those requirements that are not covered by existing Polish standards, for each class. During the review, the basic stages of the system lifecycle were considered:

1. Installation;
2. Commissioning;
3. Operation (including maintenance);
4. Decommissioning.

The following table provides an example of the identified requirements not covered by existing standards for all project options, and the benefits.

| Requirement Type Benefiting area | Project option | Requirement/ Benefit description | Comment |
|---|---|---|---|
| Environment | Options 1 & 2 | Noise management. Average level of noise, throughout the day, and during specific periods (day-time, night-time), should be kept within allowed limits. Possible approaches are:<br>1. Screens<br>2. Restricted use area procedure<br>3. Speed limit | Level of noise is likely to increase over time, if the infrastructure or rolling stock are not maintained correctly. |
| RAM | Options 1 & 2 | Point heating needs to be adequate for speed and frequency of traffic | |

| | | | |
|---|---|---|---|
| Safety | Options 1 & 2 | Project should consider/analyse EMI issues in consideration of supply of electricity to neighbours. | |
| Environment | Option 1 | Minimising risk of ground and surface water contamination by hazardous goods in regular operation. | |
| RAM | Options 1 & 2 | Improved technical condition of structures. | |
| Commercial | Option 2 | Adjustment of whole line for 200km/h for option 2, leads to significantly reduced travel time | Only applies to E20 - not CE20. |
| Safety | Options 1 & 2 | Improved safety at road/rail crossings | |

**Table 4-12: Example of high level safety requirements**

## 4.5.5 Identification of Hazards

This analysis has been based on the above system description. In support of this analysis, the author prepared a list of 32 high level railway hazards. A sample from the list of hazards given in the Table 4-13 below was reviewed by the team of experts to establish its completeness.

| Ref. | Hazard name | Hazard Description/Scope |
|---|---|---|
| CH1. | Abnormal deceleration | The hazard “Abnormal deceleration” includes only those instances of a train’s slowing sharply when not actually part of a derailment or collision scenario. Includes all 3 groups, Passengers, Neighbours and Workers. |

| | | |
|---|---|---|
| CH2. | Contaminated water and/or land and/or air | The Hazard for Contaminated Water and/or Land has been defined as the release of harmful substances likely to cause contamination of the environment. This allows the consideration of detection, mitigation and remediation barriers in the consequence domain. The release of toxic gases likely to cause harm to workers or neighbours has also been considered under this core hazard. This Core Hazard considers harm to workers or neighbours as a result of coming into contact with land, water or air contaminated with harmful substances, rather than coming into contact with the harmful substances themselves - although the toxicology is similar, the frequency and dispersion will differ. Includes Neighbours and Workers. |
| CH3. | Crossing running railway at level crossing | Includes all situations in which a user (i.e. a Neighbour) is present on a level crossing without the intended degree of protection from trains. This may arise from intentional or inadvertent misuse of the crossing by the Neighbour as well as from failures and errors in railway equipment and procedures.<br>The definition excludes situations in which harm may arise when using a level crossing as intended, for example if a user falls and injures themselves on a crossing but is still able to cross within the design time limit. The hazard excludes incidents at level crossings resulting from suicide or attempted suicide - these are assumed to be covered under Abnormal or Criminal Behaviour<br>The hazard is limited to Neighbour hazards and thus does not consider hazards at worker crossings provided within stations, depots, sidings etc. Unauthorised Neighbour use of such crossings should be regarded as Abnormal or Criminal Behaviour, being a form of trespass. (Unauthorised passenger use is covered in Hazard Inappropriate Separation between Running Railway and people.) |

| | | |
|---|---|---|
| CH4. | Electro-Magnetic Interference (EMI) caused to by railway operations | EMI Caused by Railway Operations to Businesses, General Public, Adjacent Buildings, Hospitals, has been developed to include those situations where EMI from the infrastructure or rolling stock could affect the safety of neighbours directly. This hazard does not include EMI caused by infrastructure or rolling stock to signalling and track circuits, or interference between the rolling stock and infrastructure. Such interference could be considered part of the causes for other hazards. |
| CH5. | Inappropriate separation between trains | The hazard Inappropriate separation between trains includes the scenarios in which the determined separation between trains, normally provided by the signalling system, has broken down. This hazard is defined such that there is no interface between it and the "Loss of Balance" hazard.<br>Includes all 3 groups, Passengers, Neighbours and Workers. |
| CH6. | Inappropriate working methods/environment | The scope of this hazard was defined to include most "occupational" accidents where typically a single worker is affected. Also included is the case of crane loads and other mechanical equipment fouling trains passing nearby as this was always due to operator error. |

**Table 4-13: High level railway hazards**

Each identified interface was reviewed against a list of hazards and related to applicable hazards as in a Table 4-14 below.

| Related interfaces | Related core hazards |
|---|---|
| I088: Civil Structures - PWay | CH27. Unsound/unsecured structures;<br>CH15. Incompatibility of train and structure gauge. |
| I089: PWay - Drainage | CH27. Unsound/unsecured structures;<br>CH03. Contaminated water and/or land and/or air;<br>CH18. Loss of train guidance (derailment);<br>CH30. Loss or degraded train detection. |

| Related interfaces | Related core hazards |
|---|---|
| I084: PWay - Traction Power Supply | CH06. Electro-Magnetic Interference (EMI) caused to by railway operations;<br>CH13. Inappropriate separation between uninsulated live conductors & people. |
| I085: PWay - Overhead line | CH13. Inappropriate separation between uninsulated live conductors & people;<br>CH15. Incompatibility of train and structure gauge. |
| I086: PWay - Trains | CH09. Impact from railway construction/maintenance works;<br>CH20. Objects/Animals on the line;<br>CH18. Loss of train guidance (derailment);<br>CH02. Abnormal deceleration;<br>CH31. Loss of adhesion (abnormal acceleration/deceleration). |

**Table 4-14: Initial System Hazard Log sample (Core Hazards referenced to originating system interfaces)**

The result of this study established a hazard log for the project and was used to evaluate different project options, before choosing the most appropriate solution for implementation within the funding scope and proposal for modernisation of the Corridor II.

## 4.6 Chapter Conclusions

In this chapter, the author depicted the experience and findings related to an application of some of the existing methodologies on four very different railway projects as well as the background to, and results of, the implementation of several innovative methodologies that the author developed whilst working on these projects.

A more detailed critique of the methods detailed here is provided in the next chapter.

*The safest road to hell is the gradual one - the gentle slope, soft underfoot, without sudden turnings, without milestones, without signposts.*

*—C.S. Lewis*

*“Any intelligent fool can make things bigger, more complex, and more violent, It takes a touch of genius – and a lot of courage – to move in the opposite direction.”*

*—Albert Einstein*

# CHAPTER FIVE

# CRITIQUE OF CURRENT THINKING, AVAILABLE TOOLS AND PRACTICE: A RESEARCH AGENDA

## 5.1 Chapter Introduction

The requirements for the new system based framework for system safety analysis and management are identified and outlined in this chapter. Based on that existing tools and methodologies have been assessed for compliance with requirements and a research agenda set out.

## 5.2 Summary of Requirements for new framework

The following generic safety risk analysis and management stages applicable to any system (or undertaking) have been identified as part of this research:

1. System Conceptualisation, Representation and Scoping (System Analysis). This stage of the analysis is often omitted from safety literature and standards. This preparatory phase is necessary in order to provide a structured framework and systematic approach for the hazard identification, risk assessment, and for supporting a holistic approach to the analysis. Some form of system description model, for example state transition model or sequence and collaboration diagrams, should be used as the basis for hazard identification, as the hazards resulting from each system interface, process or interaction can be elicited. The novel approach, developed as part of the research, to system conceptualisation in support of safety analysis, is discussed later in the book;

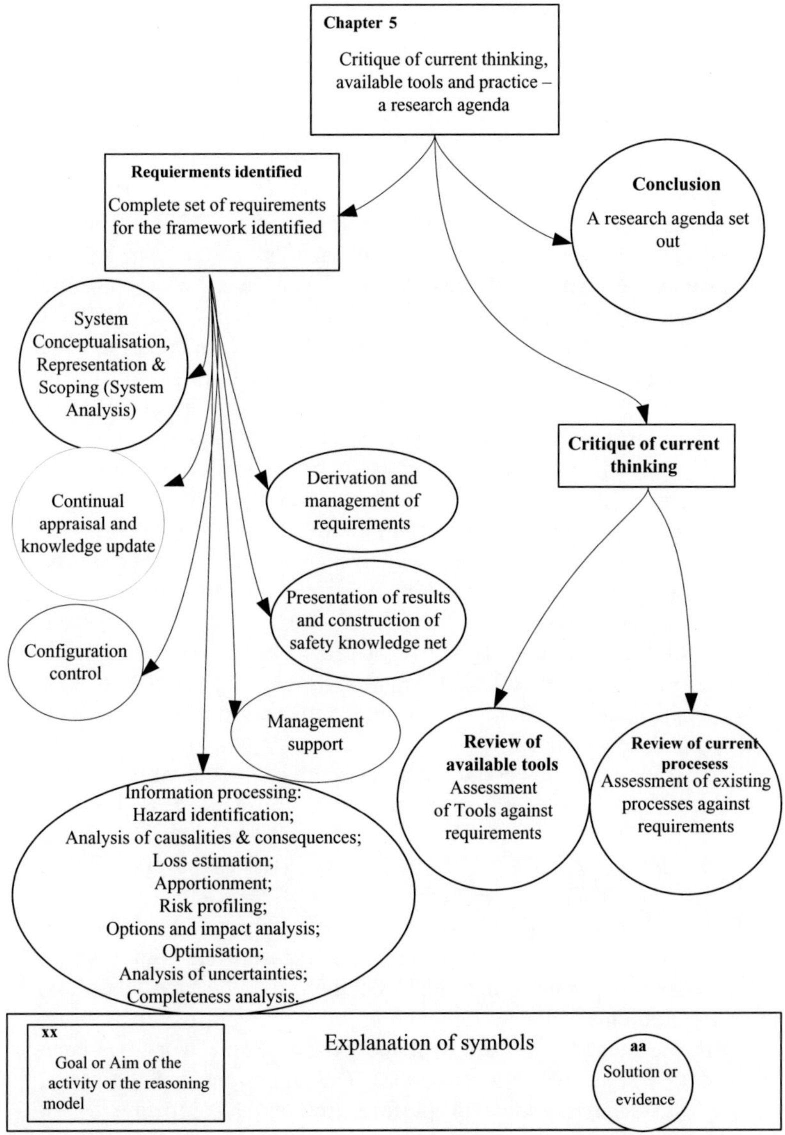


Figure 5-1: Structure of chapter 5

2. Information processing can be further structured in a number of activities:
   a. Hazard identification. The purpose of this stage of the process is to identify dangerous states associated with the system and its operation, as detailed earlier (BS EN 50129, 2003), (CENELEC 1998), (CENELEC 2003), (BS EN 50126: 1999), (BS EN 50129: 2003), (BS EN 61508: 2003), (RSSB, 2007). Although hazards, and the identification of hazards, are at heart of any safety analysis, not much work has been done on defining the properties of hazards. A formal definition of a hazard, as well the properties of the hazard, has been developed, and is presented later in this book. Also, in combination with the novel approach to system conceptualisation, a formal process supporting hazard identification, has been developed;
   b. Analysis of causalities and consequences. The objective of the causal analysis is to determine the relevant factors, which act singularly or in combination, in the realisation of a hazard. The objective of the consequence analysis is to identify the scenarios arising from a hazard and to map them to the consequences themselves (BS EN 50129, 2003), (CENELEC 1998), (CENELEC 2003), (BS EN 50126: 1999), (BS EN 50129: 2003), (BS EN 61508: 2003), (RSSB, 2007). As part of hazard identification, a structure is provided in support of the cause and consequence analysis;
   c. Loss estimation is about the assessment of the magnitude of the detrimental impact associated with consequences (BS EN 50129, 2003), (CENELEC 1998), (CENELEC 2003), (BS EN 50126: 1999), (BS EN 50129: 2003), (BS EN 61508: 2003), (RSSB, 2007);
   d. Estimation and apportionment of the safety risk to the source causes and failures of preventative measures is important in order to enable the identification of the most critical constituents of the system, and to focus the improvement efforts in the most fruitful areas of the system. Some work has been done in this area (Saltelli, 2002). The author developed a novel, formal methodology to perform the apportionment of risk within the cause-consequence modelling environment (Lucic, 2004a). This novel methodology is discussed further later in the book;
   e. Risk profiling is a multi-dimensional presentation of forecasts for future accidents for a system. Additional dimensions that

could be introduced may be time, space, or some other important variable parameters. A novel approach to risk profiling has been developed, including supporting tools (Lucic, 2003a), (Lucic, 2004a) and is discussed later in the book;

f. Identification of different solutions and mitigation measures (options and impact analysis) is a prerequisite to any claim that the safety risks of the system are As Low As Reasonably Practicable. This involves the identification of any additional potential mitigation measures, or different options to a system solution, or implementation that could further reduce the safety risk (Health and Safety Executive, 2009), (RSSB, 2007). A structured approach to conceptualisation and hazard identification was developed as part of the research that supports this stage of the analysis in a novel way, as discussed later in this book;
g. Optimisation of the system and selection of the most gainful options and mitigations is carried out in order to ensure the safest possible solution (Health and Safety Executive, 2009), (RSSB, 2007). This can be done using either a quantitative or qualitative methodology. Both methodologies are further developed and discussed in this book;
h. Analysis of uncertainties includes assessment of the level of imprecision of numerical data or understanding of the system composition, behaviour, and an appraisal of the potential impact of this uncertainty on the analysis;
i. Completeness analysis is necessary in order to confirm that none of the possible system's states or configurations, or potential scenarios, is overlooked. As part of a novel system conceptualisation approach, a framework for a completeness assessment and assurance has been development;

3. Derivation and management of requirements. Outputs of the analysis process are identified mitigations or options to be implemented in the system, or identified limits of a safe operational envelope for the system. These need to be expressed as requirements and managed throughout the life of the systems. Confirmation that the requirements are implemented is usually obtained through the verification and validation process (Blanchard and Fabrycky, 1998). Derivation and management of safety requirements is important, as one of the options to confirm final safety performance of the system is through the verification and validation of safety requirements;

4. Presentation of the results and the construction of a 'safety knowledge' net. Throughout the analysis, knowledge about the system is collected in a number of cycles (preliminary analysis, design analysis, system change analysis, etc.) corresponding to the principal engineering safety management activities mentioned earlier. As the development of the system is progressing, more knowledge is gathered and merged with the already processed information. This accumulated knowledge is then manipulated in order to generate the forecasts for the system's safety performance.
   Often, the complexity of the analysed system dictates that the problem domain is broken into digestible chunks. Prior to analysis, the system is broken into subsystems, elements of the big puzzle that can be easily comprehended. The analytical work is performed on these palatable chunks in order to produce constituents of the concluding result. Finally, using the elements of the answer, the freshly formed cells of knowledge are merged into a net defining the new understanding of the analysed system. This approach enables the easy update of the understanding of the system with new information.

   One of major aims of risk assessment is to ascertain the safety performance of the system and determine if it is satisfactory. Decisions about the adequacy of the safety performance of the system are based on knowledge gathered during the analysis process. The structuring of safety arguments into knowledge nets enables a systematic review of the gathered knowledge. Furthermore, it is necessary to create different views of the information; a person responsible for the implementation of a mitigation measure may not need the same information as the tester of the same mitigation, or the maintainer of the subsystem that provides the mitigation. It is, therefore, necessary to identify these different viewpoints on the knowledge net, and make it possible to create the required output (Kelly and Weaver, 2004), (Kelly, 2001). A novel process developed as part of the research, and presented later in this book, that extends the existing knowledge management and presentation methods by the use of Goal Structuring Notation (Kelly and Weaver, 2004), (Kelly, 2001);
5. Management support. Any safety management process must be capable of supporting management, and of recording any activities related to the analysis or implementation of mitigations, options and requirements, verification and validation and finally testing (RSSB, 2007), (Hessami, 1999b). Processes supporting the

management needs are embedded in the novel process developed as part of the research and presented later in this book;

6. Configuration control. At all points in the life cycle of the project, configuration control of all of the information must be kept. All assumptions must be documented and the work must be transparent and understandable to others (RSSB, 2007);
7. Continual appraisal and knowledge update. As the system is being developed, used and decommissioned, it is necessary to monitor the safety performance of the system, and to update the knowledge base as required. It is only if the knowledge about the system is up to date that the safety management of the system can be effective (RSSB, 2007). A novel process developed as part of the research and presented later in this book also facilitates appraisal and knowledge update and management.

Any new framework for the analysis and the management of system safety needs to include all of the above stages integrated as a holistic methodology.

In addition to the above, the methodology must be able to cope with the added complexities of staged project implementation, intricate supply chains and delivery organisations pertinent to so many contemporary projects.

## 5.3 Critique of current thinking

The work done under the general umbrella of safety management is in fact the elicitation and gathering of knowledge and corrective action management.

Information is collected, and presented in a hopefully easy to understand form; more information is collected and merged with already processed information, accumulated knowledge is manipulated, and the arguments in support of derivation of a safety case are generated.

Once the initial phase of the project is completed, the stakeholder should continue with the maintenance of the knowledge base and the management of corrective actions until decommissioning and disposal.

All the time, configuration control of all the information must be kept, all the assumptions must be documented and the work must be transparent and understandable to others.

From the start to the end of the process, the complexity of modern systems (engineering and organisational) necessitates the need to break the problem area into digestible chunks. Prior to analysis, the system is broken into subsystems, elements of the big puzzle that can then be comprehended. The analytical work is performed on these palatable chunks to produce constituents of the concluding result.

Finally, using the elements of the answer, the freshly formed cells of knowledge, “freshly trained neurons”, are merged into a large “brain” containing new understanding of the analysed/managed system.

The development of models in support of this process has the potential to greatly improve the development of the conclusions that underpin the safety argument. The individual projects can be supported by models as follows:

1. Models are an instrument that enables better understanding of a system being implemented/maintained;
2. The development of the comprehensive set of safety arguments would benefit from a good set of models describing the system and supporting the risk assessment;
3. A safety risk based, quantified comparison of the effectiveness of the different solutions in support of options analyses;
4. A quantified estimation of safety risk introduced to the system by implementation of new technology and processes, in support of the ALARP argument;
5. A model of the logical structures can be used in support of systems analysis thus enhancing the understanding of the technological and procedural solution for the project.

The use of modelling on a project is not a prerequisite for a successful conclusion of the undertaking, but it supports efficient, concise and structured knowledge gathering, structuring of the argument, and enables quantified support to the decision-making process.

However, modelling of any kind is expensive, and the Systems Safety Analysis and Management budget should be carefully balanced to enable completion of all steps of the process. If a decision has been made to develop models in support of the project, the complexity brought about by increased sophistication in modelling should be balanced against the increased difficulties in interpreting the model and the likelihood of introducing systematic errors. As already mentioned earlier in “Objectives and Aim of the Research“, a number of generic activities of safety risk analysis and management process have been identified that ease the process.

The table below provides an overview of these activities in relation to the methodologies reviewed in the previous sections of the book. Although all of the above mentioned methodologies are useful on their own, none of these covers the complete set of activities required to support the system safety risk analysis and management process, as depicted above.

At the moment, an integrated framework for the system safety risk analysis and management does not exist. Currently, the most advanced and structured process for Systems Safety Analysis and Management is presented by the 7 Stage Process and its extension 'Risk and Opportunity' paradigm.

The ISAE8 (a tool set used for development of parametric models described later in the book) is very comprehensive as it brings together the following facilities:

1. Hazard log;
2. Risk modelling environment;
3. Reporting.

However, most of the other generic activities are not part of the process or the tool. Most of the methodologies listed above are not particularly user friendly, and are very difficult and expensive to implement and maintain on the large scale project. As the complexity of the analysed system increases, the models produced using the existing techniques tend to increase in complexity at an even faster rate, resulting in extremely complex models that are difficult to validate and test. Finally, due to their complexity, these end up not being used in support of the projects.

Although some of the models are capable of supporting the analysis of temporal and spatial aspects of the systems safety risk, the results are snapshots in time and do not lend themselves well to the profiling of risk. Furthermore, apart from the Weighted Factors Analysis, none of the methodologies are capable of taking into account the effects of both detrimental and beneficial facets of the systems behaviour towards safety.

Development of the methodology supporting the generic activities of safety risk analysis and management process for any system would bring benefits to the industry. Moreover, since the current practice in the UK is that demonstration of compliance with the ALARP principle is structured in the form of a Safety Case, the process also needs to support the development of the Safety Case.

| Methodologies / Activities | System Description Models | System Dynamics | Failure Mode and Effects Analysis | Theories of Probability | Monte Carlo Simulation Models | Theory of Evidence | Reliability Block Diagrams and Network Models | Fault Trees – Event Trees | Markov Models | Petri Nets | Theory of Fuzzy Sets | Weighted factors Analysis | Advanced Cause Consequence Models | Parametric Advanced Cause Consequence Models |
|---|---|---|---|---|---|---|---|---|---|---|---|---|---|---|
| System analysis & conceptualisation | ✓ | na | na | na | na | na | ✓ | na | na | na | na | na | na | na |
| Hazard identification | ✓ | na | na | na | na | na | ✓ | na | na | na | na | na | na | na |
| Analysis of causalities | na | na | ✓ | ✓ | na | na | ✓ | ✓ | ✓ | ✓ | ✓ | ✓ | ✓ | ✓ |
| Analysis of consequences | na | na | ✓ | ✓ | na | na | ✓ | ✓ | ✓ | ✓ | ✓ | ✓ | ✓ | ✓ |
| Loss estimation | na | ✓ | ✓ | ✓ | ✓ | na | na | na | na | na | na | na | ✓ | ✓ |
| Apportionment | na | na | ✓ | ✓ | na | na | na | ✓ | na | na | na | na | ✓ | ✓ |
| Options analysis | na | na | na | na | na | na | ✓ | ✓ | ✓ | ✓ | ✓ | ✓ | ✓ | ✓ |
| Impact assessment | na | na | ✓ | ✓ | ✓ | na | na | ✓ | ✓ | ✓ | ✓ | na | ✓ | ✓ |
| Risk Profiling | na | na | na | na | na | na | na | na | na | na | na | na | na | ✓ |
| Optimisation | na | na | na | ✓ | ✓ | na | na | ✓ | ✓ | ✓ | ✓ | na | ✓ | ✓ |
| Construction of safety arguments network | na | na | na | na | na | ✓ | na | na | na | na | na | na | na | na |
| Requirements derivation | na | na | na | na | na | na | na | ✓ | ✓ | ✓ | ✓ | ✓ | ✓ | ✓ |
| Configuration control | na | na | na | na | na | na | na | na | na | na | na | na | na | na |
| Management support | na | na | na | na | na | na | na | na | na | na | na | na | na | na |
| Continual appraisal & knowledge update | na | na | na | ✓ | na | na | na | na | na | na | na | na | na | na |
| Reporting. | na | na | na | na | na | na | na | na | na | na | na | na | n/a | n/a |

**Table 5-1: Comparison of facilities provided by different methodologies.**

## 5.4 Chapter Conclusions

In this chapter, the author identified and outlined the requirements for the new system based framework for system safety analysis and management. Furthermore, using the framework requirements as the baseline, the author analysed the existing methodologies, drawing on the experience depicted in Chapter 4, and summarised the critique of current thinking in this chapter.

*Our plans miscarry because they have no aim. When a man does not know which harbour he is making for, no wind is the right wind.*

*—Senecca*

# CHAPTER SIX

# A NEW SYSTEMS BASED APPROACH TO SYSTEM SAFETY RISK ANALYSIS AND MANAGEMENT

## 6.1 Chapter introduction

The findings of the initial part of the research, and the development of a new system based approach to safety risk analysis and management theory, is discussed here.

Before one can begin defining the process for information gathering, conceptualisation and representation (System Analysis) in relation to safety risk analysis, assessment and management, it is necessary to define the properties of the hazard, as an element of the risk, and to outline the requirements and structure of the Safety Case as a final deliverable of the whole analysis process.

Hence, the first two sections of the chapter present the research findings with regards to this subject.

A discussion about the development of the new framework follows in the rest of the sections in this chapter. To close the chapter, a discussion is presented that integrates the novel methodologies discussed in Chapter 4, and some of the existing methods outlined in Chapters 3 and 4, into a new framework.

Figure 6-1 below provides an overview of the structure of this chapter, presents the lines of reasoning followed in support of the development, and indicates which parts of the process (or which methodology) is newly developed, or is a new application of an existing one.

## 6.2 The Hazard

A hazard is characterised by its potential to cause an accident, as already discussed earlier. However, the real picture is more complex.

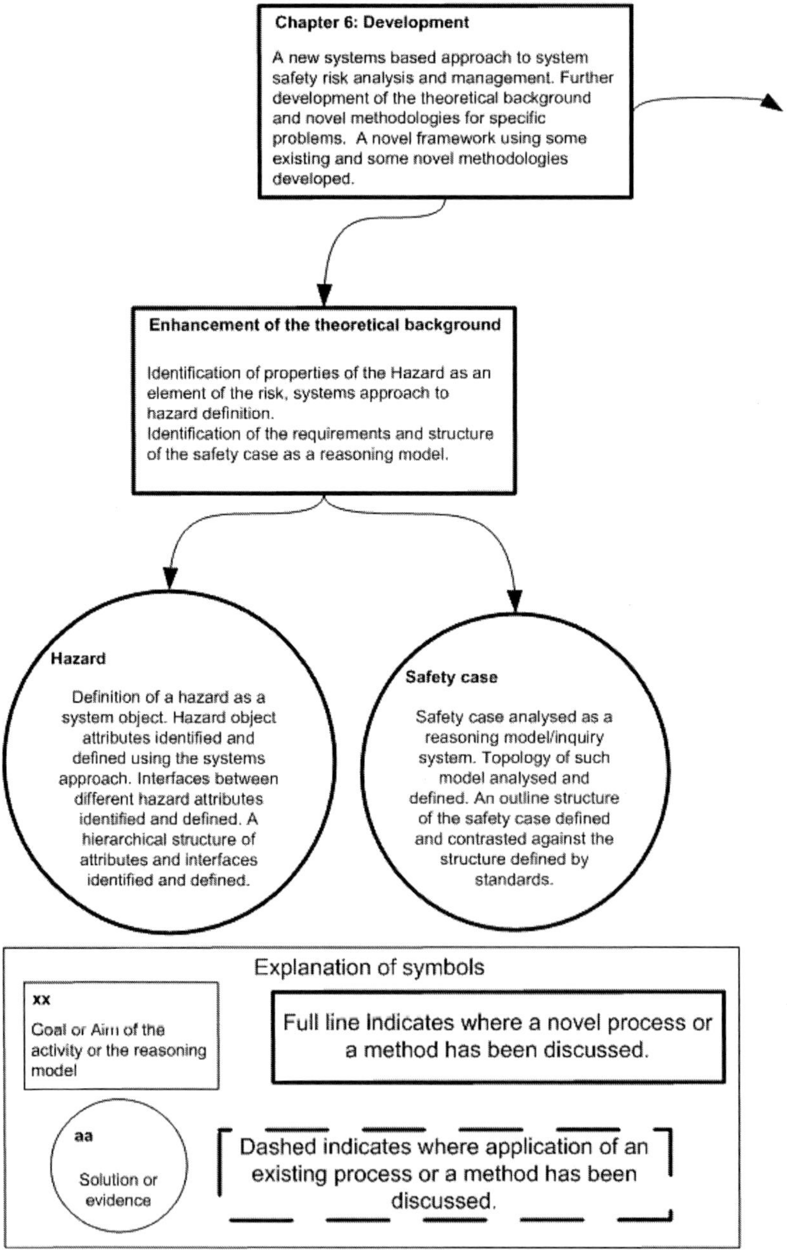

**Chapter 6: Development**

A new systems based approach to system safety risk analysis and management. Further development of the theoretical background and novel methodologies for specific problems. A novel framework using some existing and some novel methodologies developed.

**Enhancement of the theoretical background**

Identification of properties of the Hazard as an element of the risk, systems approach to hazard definition.
Identification of the requirements and structure of the safety case as a reasoning model.

**Hazard**

Definition of a hazard as a system object. Hazard object attributes identified and defined using the systems approach. Interfaces between different hazard attributes identified and defined. A hierarchical structure of attributes and interfaces identified and defined.

**Safety case**

Safety case analysed as a reasoning model/inquiry system. Topology of such model analysed and defined. An outline structure of the safety case defined and contrasted against the structure defined by standards.

Explanation of symbols

**xx**

Goal or Aim of the activity or the reasoning model

Full line indicates where a novel process or a method has been discussed.

**aa**

Solution or evidence

Dashed indicates where application of an existing process or a method has been discussed.

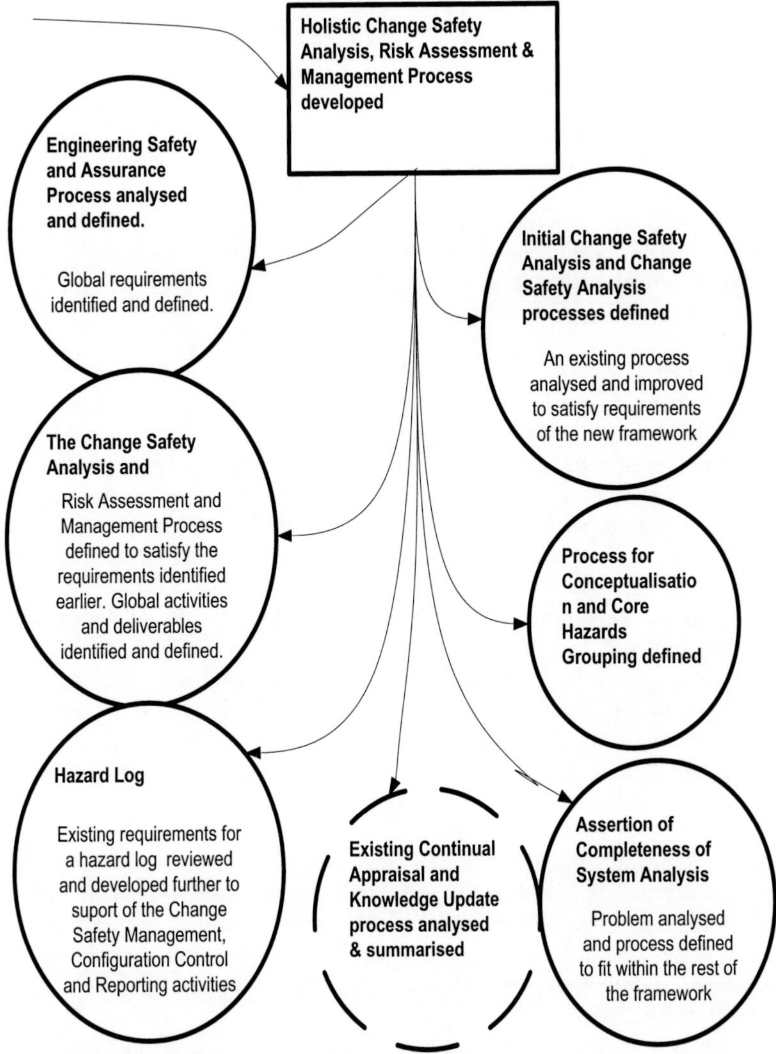


Figure 6-1: Structure of chapter 6 and the argument in support of the claim that a newly developed process is sound and complete

Different sets of attributes define different objects that make up each hazard, as well as the relationships between each individual hazard and all of the different hazards that are parts of the entirety of "safety performance" as an emergent property of any system under observation.

Furthermore, each of these sets of attributes and their internal and external relationships are characterised by their own properties. We will refer to this entirety as the "hazard universe". It is possible to identify at least two different (but related) view points on the problem domain:

1. Analytical or System Safety point of view;
2. Management point of view.

Both of these will be analysed separately and a unified, holistic relational data structure depicting the "hazard universe" will be proposed. Analysing the hazard, from a systems safety point of view, it is possible to define the "hazard universe" as a hierarchical net, consisting of a number of different objects, all of which interact to define the Safety Performance as an emerging property. If we analyse the hazard from a systems point of view, it is possible to distinguish between three sets of attributes related to each hazard:

1. Attributes related to the origin of the hazard, causality of the hazard (spatial and temporal);
2. Attributes related to the evolution of the hazard, post hazard horizon, (spatial and temporal) into a consequence or set of consequences, defining the frequency of the consequences;
3. Spatial and temporal attributes related to the consequences, defining the nature of the consequences and subsequent loss.

Interactions of these three sets of attributes then define the attributes of the hazard itself. The collective effect of interactions between different hazards (in fact, these are interactions between the attributes of individual hazards) could be seen as emerging properties of the "hazard universe", where the "hazard universe" is treated as a system:

1. Potential detriment, risk, which the hazard can initiate (as an emerging property defined by the combined of spatial and temporal effects of attributes related to the origin of the hazard, attributes related to the evolution of the hazard into a consequence or a set of consequences, consequences themselves and of any containments of consequences);
2. The spatial and temporal nature of the hazard itself (as an emerging property defined by the combined spatial and temporal effects of attributes related to the origin of the hazard).

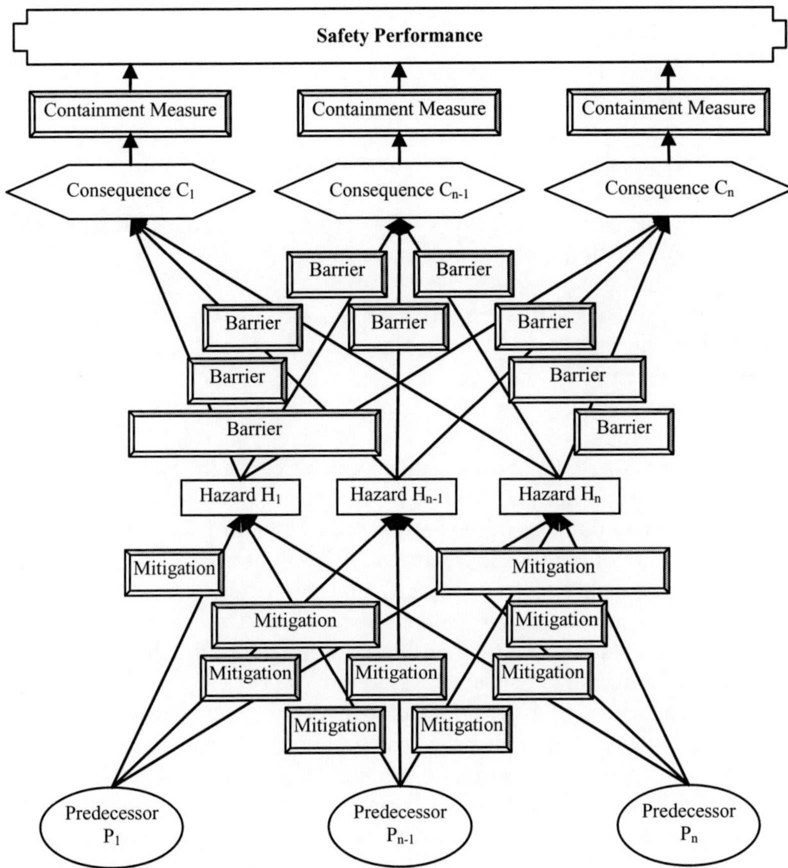


Figure 6-2: "Hazard Universe"

Attributes related to the origin of the hazard could be treated as constituents of the "hazard universe" system, as in Figure 6-2 above, which, when interacting together, define the causal properties of each of the hazards, and of the "hazard universe". Such properties are:

1. The cause or causes (predecessor(s)) of the hazard;
    a. Spatial properties, location of the predecessor(s) to the hazard;
    b. Temporal nature of the predecessor(s) to the hazard;
    c. Nature of relationship between different causes;
    d. Potential relationship between causes and protection/defence measures (causes instigating failures of protection/defence

measures);
   e. In case that there is more than one predecessor, a measure of contribution of the predecessor to the hazard;
2. The mitigations against causes. The attributes of mitigations are:
   a. Nature of relationship between the cause and the mitigation;
   b. Nature of relationship between different mitigations;
   c. A measure of reduction of the detrimental effect of the cause or contribution to the hazard.

Attributes related to the evolution of the hazard into a consequence, "post hazard horizon", could be treated as constituents of the "hazard universe" system, interacting together with which, defines the consequences of each of the hazards, and of the "hazard universe", with their own properties:

1. The barriers to the hazard, reducing the effect of the hazard;
   a. Spatial properties, location of the barrier(s) to the hazard;
   b. Temporal nature of the barrier(s) to the hazard;
   c. Nature of relationship between different barriers;
   d. In case that there is more than one barrier, a measure of the reduction of the detrimental effect of the hazard;
   e. The nature of the relationship between the causal domain of the hazard and the barriers;
   f. The nature of the relationship between different barriers.
2. Attributes related to the progression of consequences and effect of the containment measures, "post consequence horizon", could be treated as constituents of the "hazard universe" system, which, interacting together, characterize the safety performance of the system under observation and its "hazard" universe:
   a. Frequency or probability of consequence occurrence;
   b. Temporal nature of the consequence;
   c. Spatial properties, location of the consequence;
   d. Nature of subsequent loss;
   e. Exposed party (who will suffer the loss);
   f. Potential for direct consequences causing further subsequent consequences;
   g. Nature of the relationship between different consequences;
3. And finally, containment measures:
   a. Spatial properties, location of the containment measure(s);
   b. Temporal nature of the containment measure(s);
   c. Nature of relationship between different containment measures;
   d. A measure of reduction of the loss;

e. Nature of relationship between different containment measures.

In support of the safety management process, it is usual to create a log of all the hazards on the project and then to track the implementation of any defence/protection measures using this log. The author will talk about the use of the Hazard Log and its place in the management processes later. However, for the purpose of this analysis the author has identified that additional information is required. A distinction should be made between the attributes of the "hazard universe" from a systems safety point of view, and the information needed in support of safety management process.

It is possible to classify these information items into two categories:

1. Support to a comprehensive audit trial;
   a. Source of identification of each object of the "hazard universe";
   b. Any documents that are relevant/related to the object;
   c. In case of actions, information supporting closure, confirmation that the defence/protection measure is in place and working;
   d. Journal, a record of all changes to the hazard log;
   e. Configuration control information;
2. Management of actions to implement or confirm implementation of identified defence/protection measures
   a. Hazard:
      i. Status of the hazard:
         - Hazard is OPEN if the risk emerging from the hazard is unacceptable and defence/protection measures introduced or identified are not sufficient to mitigate the risk to an acceptable level;
         - Hazard is CONDITIONALLY CLOSED if the existing and new identified (but not yet implemented) defence/protection measures are sufficient to mitigate the risk to an acceptable level;
         - Hazard is CLOSED-Mitigated if the existing defence/protection measures are sufficient to mitigate the risk to an acceptable level. However it does not mean that this hazard is forgotten, as discussed earlier. We will discuss the Safety Management process later in more detail;
         - Hazard is CLOSED-Eliminated if the hazard is completely eliminated, for example by design. Again it does not mean that this hazard is forgotten and we will discuss the Safety Management process later in more detail;

  - Other statuses are possible (for example TRANSFERRED when responsibility for safety management of the hazard is transferred to some other project or party, or CANCELLED if after initial analysis it is confirmed that what was initially thought to be a hazard does not actually exist) but the first four are the most important;

  ii. Disposition statement is clarification of the hazard status and record of any changes to the hazard.

b. Defence/Protection measures:

  i. Action Description provides detail about the action to implement the defence/protection measures;

  ii. Action Status
    - Action is OPEN if it is not implemented;
    - Action is CONDITIONALLY CLOSED if it is certain that the action will be implemented;
    - Action is CLOSED if related defence/protection measure is implemented;
    - Similar to hazard, other statuses are possible (for example TRANSFERRED or CANCELLED) but unlike hazard, once the action is transferred or cancelled it is usually kept only as a record of earlier activities;

  iii. Actionee is responsible for the implementation of the defence/protection measure;

  iv. Actionee response is the status report regarding the implementation of the action, the implementation deadline date, and providing a timescale for implementation of the action;

  v. References to relevant safety requirements provides a link to the system requirements set. A subset of the defence/protection measures will refer to "things" that exist already, and some to those that are needed to make the system acceptably safe. These defence/protection measures should be transformed into requirements.

The diagram (Figure 6-3) below depicts the above described attributes (coloured yellow), information items (blue for audit trial related items and white for management of actions related items) and their relationship. Based on this structure it is possible to define different "views" on our knowledge base about the "hazard universe" under scrutiny.

The "views" will depend on the need of the viewer or interested party. As already discussed, an analyst will be interested in slightly different information than, for example, a project manager or a person responsible for implementation of the identified action or a requirement.

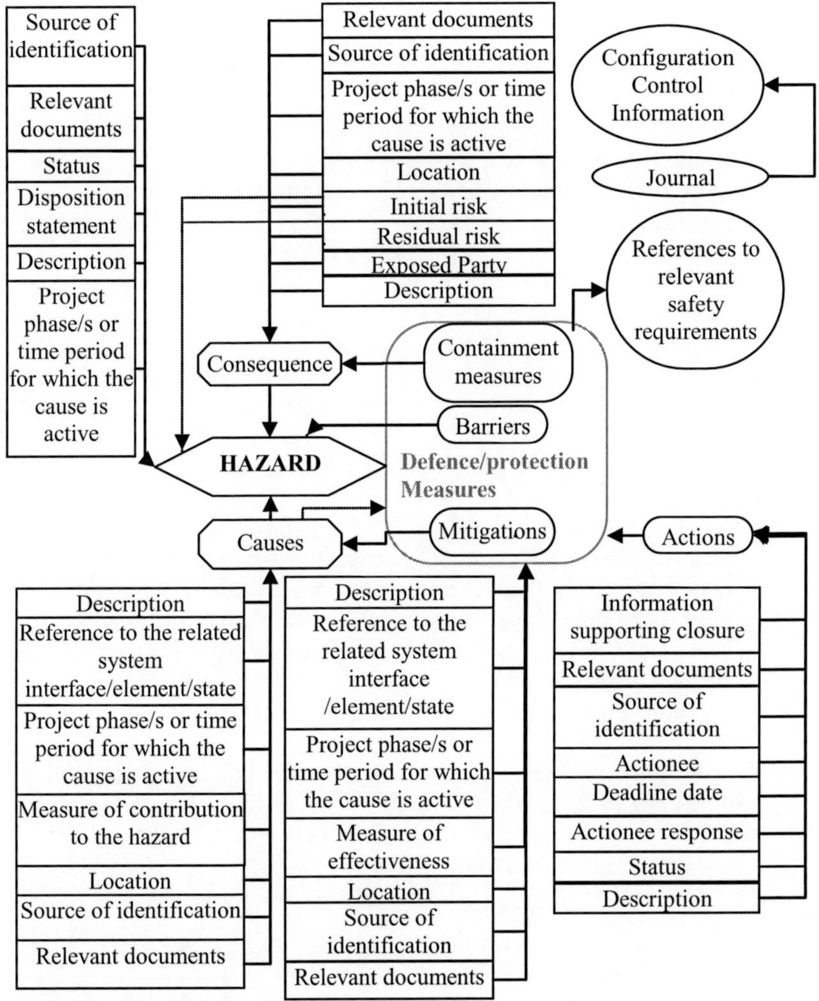


Figure 6-3: Hazard attributes

This structure is used for further analysis and development of a process for information gathering, conceptualisation and representation in support of systems safety analysis and management.

## 6.3 The Safety Case

The purpose of the safety case is to present the arguments that the system, which is the subject of the safety case, is acceptably safe to operate in a precise operational environment (Kelly and Weaver, 2004).

Each safety case consists of four elements:

1. Objectives or Requirements: The safety case must identify and present the requirements that need to be fulfilled to ensure satisfactory safety performance of the system;
2. Evidence: The Safety case needs to present the evidence that requirements are fulfilled;
3. Arguments: The safety case needs to present two types of arguments. Firstly it needs to argue that the set of identified requirements is sufficient to demonstrate, assuming that all requirements are implemented, that the system is acceptably safe. Secondly, it must argue that the evidence provided such that the requirements have been fulfilled, with sufficient proof of their requirements implementation;
4. Context: The safety case must be defined within the operational context of the system. Therefore, the safety case must identify and outline the operational environment of the system being analysed. This needs to include the technical "operational envelope" of the system operating within the given environment under all given operating conditions whether normal, degraded or emergency.

As a result of the research, the author proposes that, from a reasoning point of view, a process supporting derivation of the safety case is of a form of the Singerian inquiry. It is possible to present the Singerian inquiry system with the following model:

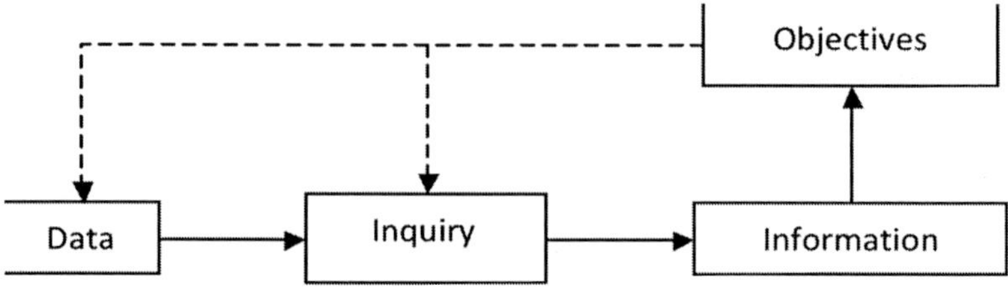


Figure 6-4: Singerian inquiry system model

In the Singerian inquiry model, the choice of objectives determines the choice of data to be used, and the structure of the holistic inquiry model.

Following the same approach as the development of the system, the system representation and identification of boundaries between subsystems/constituents of the system must support the above listed constraints. According to the Singerian inquiry model, the safety case should be structured around main safety goals and objectives of the system, and then decomposed into lower level goals and requirements.

It is possible to treat a safety case derivation process as a system in its own right. If the generalised description of a system, as mentioned earlier, is used, a special case of that system would be a “safety case derivation process system”.

The “safety case derivation process system” is a conceptual, hierarchical system, made up of information packets linked together into a fact net model, with a function to infer and present the reasoning why the undertaking, which is an object of the safety case, is sufficiently safe.

In the generalised system description, described earlier, information, energy, and action, are three general categories of inputs and outputs of the system. Also, a special class of inputs and outputs of the system related to the environment was identified. In the case of the “safety case derivation process system”, elements of a generalised system definition can be characterised as follows:

1. The system corresponds to the “safety case derivation process system”;
2. The environment corresponds to the analysed system, context within which the safety case is being developed, legal constraints and societal perceptions;
3. The system constituents correspond to a number of different “objects” as follows:
    a. The safety goal, or on a lower level of a hierarchy, the safety objective or safety requirement. The safety goal is influenced by external inputs such as the legal requirements or constraints and societal perceptions and expectations, to mention some;
    b. The data, which in case of the safety case process, relates to raw information being analysed, or to evidence. The data is taking as input information from outside the “safety case derivation process system” from across the system boundary;
    c. Inquirer, which in the safety case process terminology, relates data to the information that produce the argument;

 d. The argument, which in the safety case terminology, relates to a chain of reasoning that supports the claim that the safety goal has been achieved;
4. With regards to the interfaces of the system, three categories were identified earlier, "information", "action" and "energy". In the case of a "safety case derivation process system", since it is a conceptual system, interfaces of "energy" and "action" category do not exist as such. All of the interfaces are in fact "information" interfaces, relating arguments and evidence to safety goals, safety objectives or requirements. In the case of external interfaces these provide data to support the enquiry process and information to mould the safety objectives;
5. The topology of the system corresponds to the structure of internal and external interfaces of the "safety case derivation process system".

Extending the same principle the "safety case derivation process system" within its environment, can now be presented by Figure 6-5:

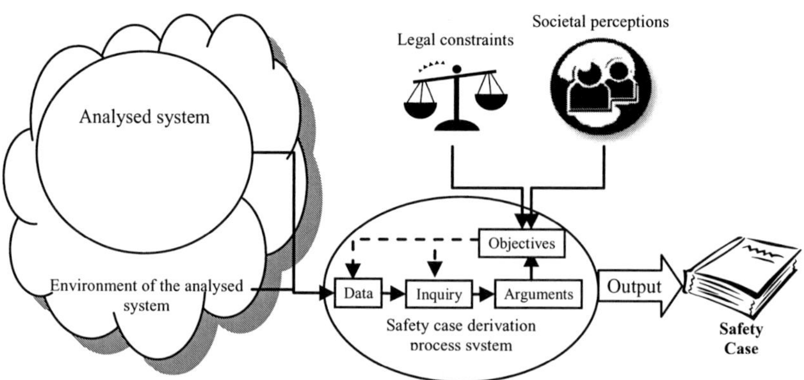


Figure 6-5: "Safety case derivation process system" within its environment

Since the "safety case derivation process system" is in fact the Singerian inquiry model, the choice of objectives will influence the choice of the inquirer and the sought after data. It is consistent with current practice that, when analysing software related systems or subsystems, different methods will be used, than if the subsystem is hardware based. Therefore, different inquirers of the "safety case derivation process system" make up a hierarchy of inquirers related by higher level inquirers and, as a result, transform into a final argument supporting the objective.

Figure 6-6 depicts the extended knowledge net that is the "safety case derivation process".

The conventional safety case development method is focused on a static system definition, wherein for any change to the system configuration, a new bespoke safety case needs to be developed (Kelly, 2001). However, modern systems and recent large scale projects introduce the additional complexity of a constantly changing system configuration and the environment within which the system is operating (the system here is understood to be the wider system including the operators and users). If a conventional approach to the development of the safety case is to be followed, it would mean that for each system configuration, or stage of the project, a new bespoke safety case would have to be produced.

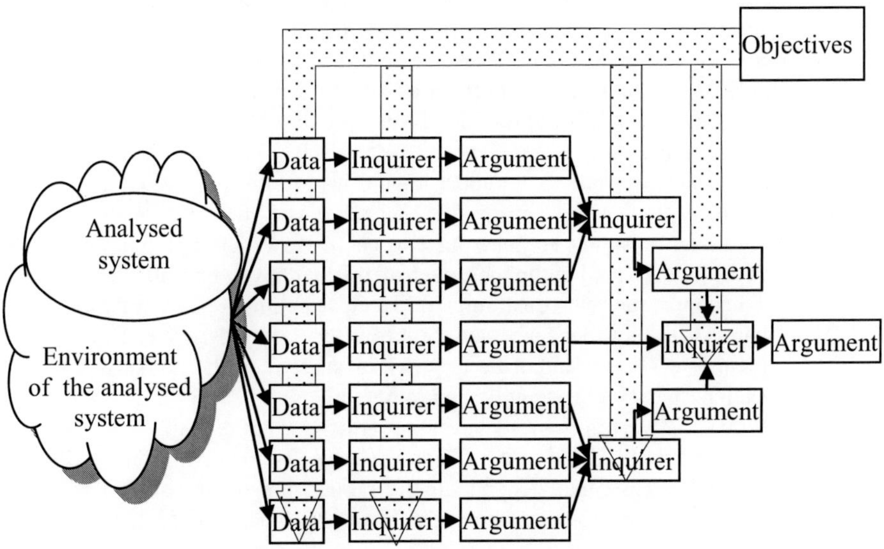


Figure 6-6: Generic safety case derivation process system

A more appropriate approach would be to develop a safety case structured into modular elements, which should enable an easy way to adopt or change the safety case in order to represent each new reality. In the interest of efficiency, economy and the effectiveness of the safety assurance process, the safety case for each system configuration or project stage, should be structured in such a way as to enable:

Constraint 1. Reuse of material and the structure of safety case supporting the previous system configuration or the preceding project stages to the maximum extent which is appropriate and feasible;

Constraint 2. For each following issue of the safety case, the underlying safety argument should build upon the safety arguments, safety processes and safety management activities as developed during the previous stages of the project or in support of earlier system configurations.

In support of this approach, the author identified two potential methods to structure the safety case "net model". Firstly, one could partition the system into parts following some predefined methodology, and structure the safety case around these parts (Kelly, 2001). This way, the final safety case would, in fact, consist of a number of lower level safety cases, one for each part of the system, brought together by an overarching argument that the overall system is safe.

This approach appears as the most logical way to define a structure of the safety case "net model", primarily because it is a direct extension of the conventional systems engineering approach. However, this approach brings about a number of complex requirements that have to be fulfilled:

1. To support the two constraints above clear boundaries, between the parts of the system to be supported by the safety case, need to be identified. There cannot be any overlaps between the parts, as for each part a safety case will need to be built[2];
2. Since the partition of the system requires a clear split between different parts of the system, it is necessary to identify all parts of the safety case that depend on the input from other parts of the safety case (mentioned above), and a decision has to be made in which part these will be included;
3. Decomposition of the analysed system also needs to consider the real system boundaries that may change throughout the life of the project or the product;
4. Since the first assumption about the analysed system is that it is a changing system and as we need to analyse the impact of the change to overall system safety performance, it is likely that this approach would necessitate complete review of the safety case each

[2] In practise this are often not referred to as safety cases, but as parts of the overall system safety case. However each one of these possesses all the attributes of a standalone safety case.

time and careful identification of the changed parts that require a new safety case.

An alternative approach is to identify a common "theme" across all different phases of an emerging project or all different manifestations of an evolving system. From early stages of the project onwards, for any system it is possible to identify the Core Hazards emerging from that system.

For example, on railways, one of the Core Hazards would be the "Inappropriate separation of trains". Furthermore, it is very rare that the system to be implemented is so novel that it is not possible to research the history or to use previous experience to identify a comprehensive list of potential Core Hazards. However, this is an iterative process and the completeness of analysis should be assured through a systematic analysis and review process as is explained in following sections of this book.

An overall Safety Goal is that for all Core Hazards, the emerging safety risks are managed to an acceptable level. Therefore, for each Core Hazard, a safety case like argument should be made to prove that the risk brought about by a particular Core Hazard is managed to an acceptable level.

This approach has several significant advantages. The "safety case derivation process system" is about proving that the emerging safety performance of the analysed system is satisfactory, and its output, the safety case, is about presenting the results of the analysis in a logical manner. Therefore, it is logical that the "safety case derivation process system" is structured around the Safety Goals and analysis of the related Core Hazards, as it is the successful management of the Core Hazards that define the level of safety risk.

This approach should also make it possible to identify the parts of the subsystem relevant to a Core Hazard, and then analyse them in the context of that particular Core Hazard. For each new manifestation of the system, a review of the system elements that contribute to the Core Hazard should be undertaken in order to identify those that are changing.

The analysis can then focus on the changes to the system elements, and the impact of these changes on the overall safety performance. In order to satisfy the two constraints described above, the structure of the "safety case derivation process system" net model, and consequently the safety case arguments, needs to allow easy identification of the parts that are changing from one system manifestation to the next and the impact of these changes on the overall system safety performance.

Goal Structuring Notation (GSN) as a presentation of the net model lends itself very well to this task. Each of the main elements of the safety case, Objective, Argument and Evidence, as noted above, is represented by a symbol. Notably, Context as a part of the safety case is not presented uniquely by a GSN symbol (Figure 6-7). The Constraint as a symbol of the GSN is reserved for any type of constraint on the system and therefore includes the Context of the system operation. Other symbols of the GSN will be discussed later in this book.

In terms of a structure and specific areas of interest to be covered by any safety case, a structure recommended in EN50129 and the Yellow book (RSSB, 2007) is widely recognised as representing a good practice in the production of safety cases and is used well beyond the standard's nominal scope of electronic systems for railway signalling.

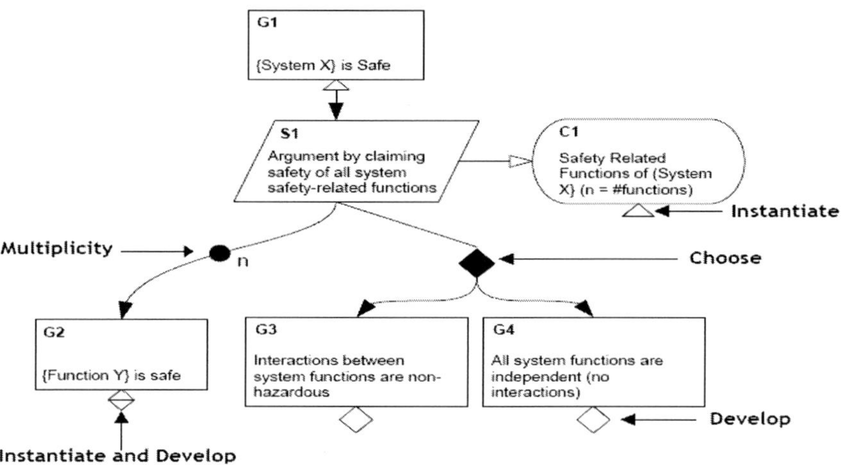


Figure 6-7: Example of reusable GSN pattern

The following are generic sections which correspond to the four elements of the safety case listed above:

1. Introduction: High level safety objectives or requirements should be presented in this section;
2. System Definition: A safety case can only "exist" within a context of the system for which the safety case is being written. This part relates to bullet point "4. Context" above;
3. Quality and Safety Management Reports: In support of the safe system operation it is important to develop and implement an adequate Quality and Safety Management Systems. This part of the safety case should present the Quality and Safety Management

System adhered to during the life cycle phases of the system to be supported by the safety case. This part of the safety case tallies with bullet point "2. Evidence" above;

4. Technical Safety Reports: This section of the safety case contains all safety reports related to the system. This section corresponds to the main safety case parts, Objectives or Requirements, Evidence and Arguments. The context of this section should not be a simple repetition of the previous two sections but should go into as much detail as is required to fully demonstrate the safety argument;
5. Related Safety Documents: This section should contain references to all other related safety documents used in support of the development of the safety case.

Later it will be show how the structure of the safety case recommended by EN50129 can be fused with the approach suggested here.

## 6.4 Holistic Change Safety Analysis, Risk Assessment and Management Process

### 6.4.1 Engineering Safety and Assurance Process – Big Picture

Change Safety Analysis and Risk Assessment Process is the part of the overall Engineering Safety and Assurance Process (ESAP). The author developed ESAP as a generic System Integration process. For clarity, firstly the ESAP will be depicted and, following on from there, details of the Change Safety Analysis and Risk Assessment Process will be provided.

It is difficult, if not impossible, to know what the outcome of a complex change to a complex system will be. The system's emergent properties can lead to novel, atypical, and dangerous failures. Furthermore, the exhaustive testing of complex systems is most often an impossible task. Hence, the only pragmatic technique to ascertain the entirety of the "hazard universe" must be centred on a systematic analysis process.

The process flow diagram in Figure 6-8 describes the information flow, for each major system configuration (related to different project implementation stages), and, at a high level, the inputs, key activities and outputs of the overall systems integration process. At the centre of the process is the System Definition that describes how the elements of the system relate to one another to provide an integrated system, metamorphosing from one system configuration to the next.

The initial requirements capture process brings together all of the known sources of safety, functional and performance system requirements, that were documented at the beginning of the project, including the original concept specifications.

Based on these requirements, the system definition is developed to capture the baseline system configuration, before any changes brought about by the project, and subsequently before each of the project stages.

The high level system definition is in the form of a physical architecture diagram depicting the high level architecture of existing systems. Underlining the system architecture, and based on the understanding of existing interfaces, the System Context and Interface Diagrams (SCID) were developed.

As the performance of the interfaces defines the performance of the system for each of the emerging properties/categories, a system is delivering against: safety, RAM and performance, the detailed information about the interfaces, including the stage of the project (system configuration) the interface is active for, is captured in the Objects (assets, procedures and people) and Interfaces Database (OIDB).

The system definition is utilised to analyse the implications of the change on the system and identify detailed requirements against the deliverables, including, Systems Safety, RAM, Human Factors, Operational Readiness, EMI/EMC and Performance.

The first step towards obtaining acceptance of the new system is the demonstration that what has been designed meets the requirements as specified during the course of development.

This is done through verification and validation. The aim of the verification is to show that a specification at one stage in the delivery life cycle is consistent and compliant with the specification of the previous stage, taking account of any agreed changes or variations.

Maintaining formal control over the traceability between the requirements repository (DOORS software package) and the project remit and standards, enables the project to prevent requirements creep, substantiate its argument for acceptance of the deliverables, and identify any changes to the system performance.

Subsequent stages of verification should be between the requirement sets and the system definitions produced at each stage of the programme/project life cycle (architectural design, system design, sub-system design and detailed design). Validation is undertaken to demonstrate that what has been built against the design meets the stated requirements.

During the course of systems development, all parties involved in the delivery of the changes under the project will be compiling evidence that the processes have been followed, that safety and performance issues are well understood, and that the system to be delivered meets all requirements and specifications.

Consent to move to each new system configuration state is sought by an "application to consent to operate" or "consent to trial operations", supported by an Engineering Safety and Assurance Case (ESAC), which sets out the justification for the assurance of the system, as it is at that stage.

The ESAC presents the arguments for the assurance of the overall system in a given major configuration as affected and delivered by the project works.

The ESAC outlines the system assurance argument by bringing together four distinct areas of assurance:

1. System Safety;
2. Operational Readiness;
3. System Reliability, Maintainability and Availability (RAM);
4. Performance.

The structure of the system safety case is discussed later.

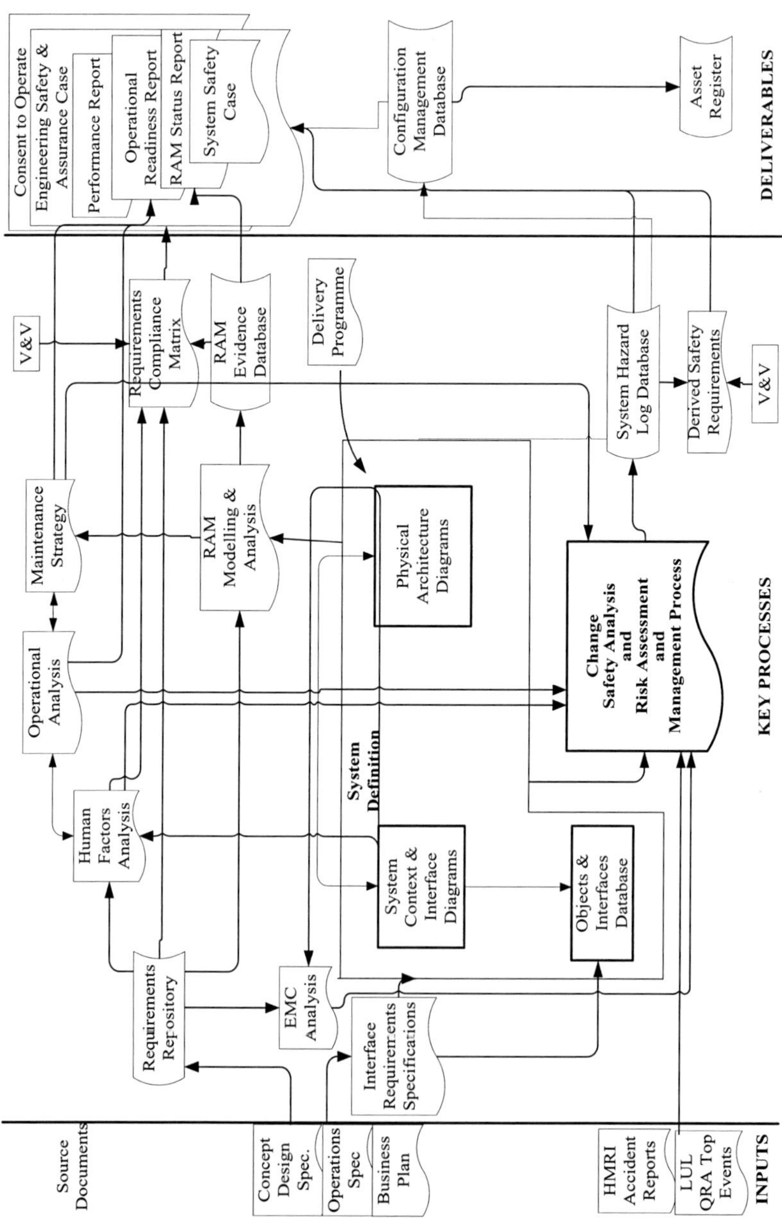


Figure 6-8: Engineering Safety and Assurance Process

## 6.4.2 The Change Safety Analysis and Risk Assessment and Management Process

The complexity of modern undertakings necessitates an integrated approach to system safety hazard management, quantification of system safety requirements and system safety assurance. The process described in this book aims to integrate these three aspects of System Safety Management. To demonstrably discharge the duty of care, it must be ensured that all changes are subject to safety analysis, that a robust process has been followed for the management of system safety hazards and that all hazards have been satisfactorily resolved. However these changes vary enormously in significance and, unless the safety analysis processes are scalable, they will either impose an unacceptable overhead on the less significant changes or fail to cope properly with the more significant ones. As a scalable change analysis process, the Change Safety Analysis and Management process (provides the framework for achievement of these goals), is extended to include quantified risk assessment modelling and systems safety requirements derivation, quantification and management. The Change Safety Analysis, Risk Assessment and Management Process encapsulates the following:

**1.** System conceptualisation and core hazards grouping;
**2.** Scoping of the impact of the change on safety performance of the system;
**3.** Safety risk analysis and assessment:
   **a.** Hazard Identification;
   **b.** Cause Analysis;
   **c.** Consequence Analysis;
   **d.** Loss Analysis;
   **e.** Safety requirements derivation;
**4.** Qualitative and Quantitative derivation of safety targets and valuation of safety requirements;
   **a.** Options and Impact analysis:
   **b.** Apportionment and risk profiling;
   c. Impact assessment and optimisation;
**5.** Management of safety requirements;
**6.** Configuration control;
**7.** Necessary outputs/reports;
**8.** Input into construction of safety arguments network;
**9.** System Safety Hazard Log management.

Figure 6-9, below, depicts the CSA, Risk Assessment and Management Process, including the hazard management process and the identified dependencies and the deliverables. The process is heterogeneous and adaptive to any system and encapsulates following activities, dependences/inputs and deliverables:

| | |
|---|---|
| 1. | The main input to the process is the System Definition. The system definition is used to scope and baseline the system, and initiate the analysis process; |
| 2. | As the undertaking is progressing, the initial requirements are subject to changes and these changes must be assessed for impact on the system definition and the system definition should be updated accordingly. Typically on a project these should be done through the Engineering Change Request (ECR); |
| 3. | Usually one of the main reasons for change of the initial requirements set is the feedback from initial testing of the system recorded in the Defect Recording and Corrective Action System (DRACAS); |
| 4. | Using the system definition information the system is conceptualised, resulting in the set of process models of the system that in turn enable further quantified and qualitative analysis; |
| 5 and 19. | The system model is then used to support the Initial Change Safety Analysis and development of the QRA. Output of this activity is the Initial Change Safety Analysis record. This record is configuration controlled, and represents one of the building blocks of the completeness of the analysis assurance; |
| 6, 18 and 23. | Those changes that are categorised as category one or two are subjected to Change Safety Analysis (CSA), all novel changes to the system are subjected to safety analysis. Therefore, like for like replacement of existing equipment is not subjected to safety analysis. Output of the CSA process is in the form of briefings, in preparation and support of the analysis and the reports detailing findings of the CSA; |
| 7, 20, 21, 22, 23, 24 and 25. | Outputs of the CSA provide input into the System Hazard Log. The scale and complexity of the project necessitates a central system hazard data store, referred to as the System Hazard Log (SHL), in support of the derivation of the safety case. The System Hazard Log is the central repository of all the data related to the CSA process, and the tool for management of the actions related to the implementation of identified mitigation measures. All hazards are entered into the SHL, and hazard closure requires |

| | |
|---|---|
| | review and endorsement at the Hazard Forum. Actions to implement requirements and mitigations/containment measures are managed and tracked through the SHL as well. From the SHL, System Hazard reports are produced in support of the System Safety Case submission, depicting the status of the CSA activities and the SHL. System Hazard Log Reports are produced quarterly, and provide detailed overview of the content and status of the SHL. A significant part of the safety arguments and evidence enabling closure of the SHL records and completion of the Systems Safety Case will come from the safety documentation produced by the supply chain. Finally, results of the System Hazard Log Implementation audits are fed to the SHL and into the System Hazard Log Implementation audits; |
| 8, 16 and 17. | System conceptualisation and the outputs of the CSA are used to support quantified or qualitative risk assessment, for each identified Core Hazard.<br>It should be noted here that a choice of the risk assessment methodology (quantitative versus qualitative) depends on the nature of the risk being analysed and the needs of the safety analysis; |
| 9. | Both, the risk assessment models (quantitative or qualitative) and the System Hazard Log are "synchronised" in terms of information and coverage of the hazards; |
| 10. | Safety requirements should be derived from the defence/protection measures stored in the SHL and some of these requirements will be quantified using the Quantified Risk Assessment model. A Safety Requirements Forum be formed, to assemble sufficient expertise necessary to elicit safety requirements; |
| 11 and 12. | Risk assessment is done at two levels, firstly overall system safety performance targets are established, and then the requirements quantified. It is also necessary to check that results from testing and monitoring systems comply with the quantified safety targets; |
| 13, 14 and 15. | A team of experts in required disciplines, independent from the project, oversee critical CSA activities, endorse the output (reports) of the process and approve the change of status of SHL actions and consequently the Core Hazards. In support of the Hazard Forum's decision making, for any change of the SHL action status a proof of implementation is sought in the form of an actionee response, V&V report, test log, etc. |

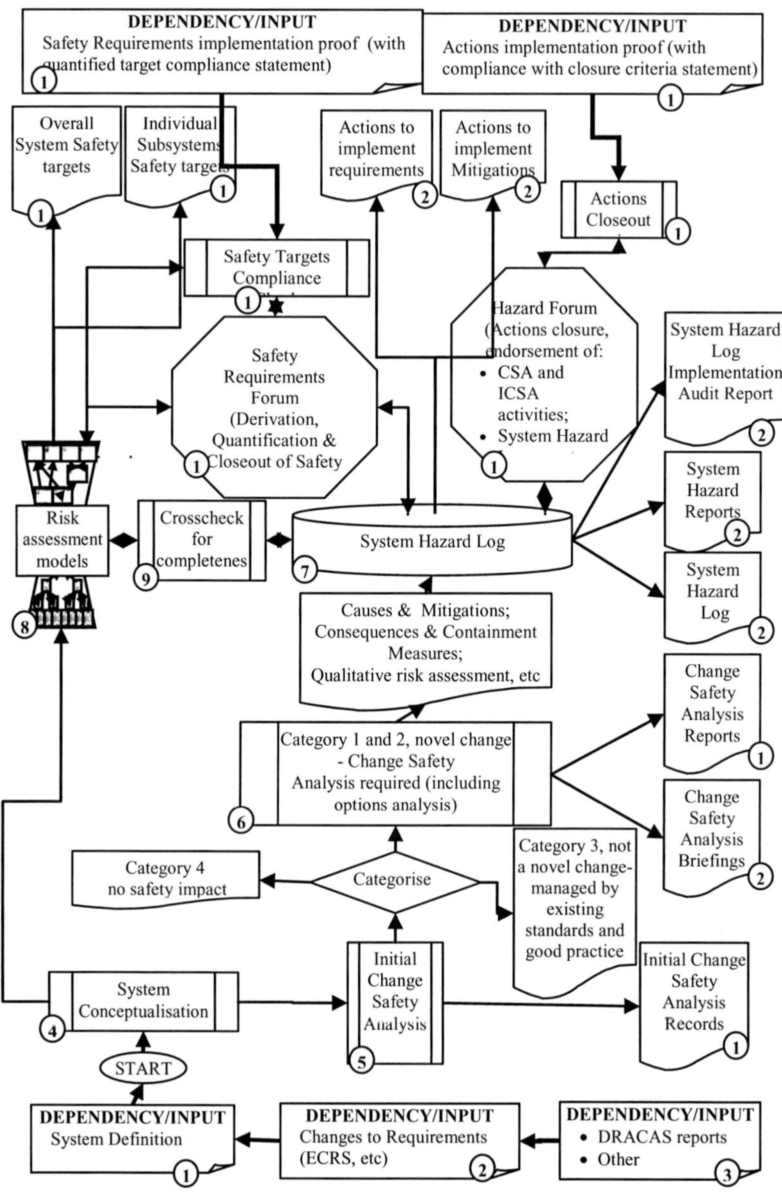


Figure 6-9: CSA and Risk Assessment Process

Detailed descriptions of each step in the process are provided in the following sections.

### 6.4.3 Conceptualisation and Core Hazards Grouping

To enable the systematic analysis of any system it is necessary to understand the whole of the system, to conceptualise the system. Earlier, several different methods of conceptual modelling were discussed, and also some of the applications for conceptual modelling were presented.

For a complex system, it is usually necessary to produce a hierarchy of conceptual models, very detailed representations of the system to capture all of the system constituents and their behaviour (these models will be referred to as System Concept Interface Diagrams (SCID)) and higher level system representations capturing the subsystems and their interactions. Still, some of the system architecture models contain hundreds of interfaces and objects, and are very difficult to analyse and keep under configuration control. For example, a System Definition for the Victoria Line Upgrade Programme (VLUP) 3, consists of three major elements:

1. Physical Architecture Diagrams, which represent the physical assets and their relationship for all stages of the project as it transforms through different system configurations;
2. System Context and Interface diagrams which model all system constituents including human interactions and the environment;
3. Objects and interface database which contains detailed description of all objects and interfaces within SCID and descriptions of changes to these for different system configurations.

In support of the conceptualisation process a new type of modelling was suggested. If one defines the "core" hazards for the analysed system, then it is possible to extract from the detailed system description (SCID), only those parts of the system that contribute to the core hazard and produce a conceptual model describing the relationships between all of the system constituents (including environmental inputs).

Through analysis of historical data and a high level system analysis, it is possible to identify the "core" hazards, and use them for the grouping of hazards and process models.

---

[3] Victoria Line Upgrade Programme was used to test the processes described in this section of the book. This work is depicted later, in chapter "Application" of the book.

The concept of the process model methodology was developed by Short of Atkins Global (Short, 2007), in support of the development of safety cases for complex railway systems.

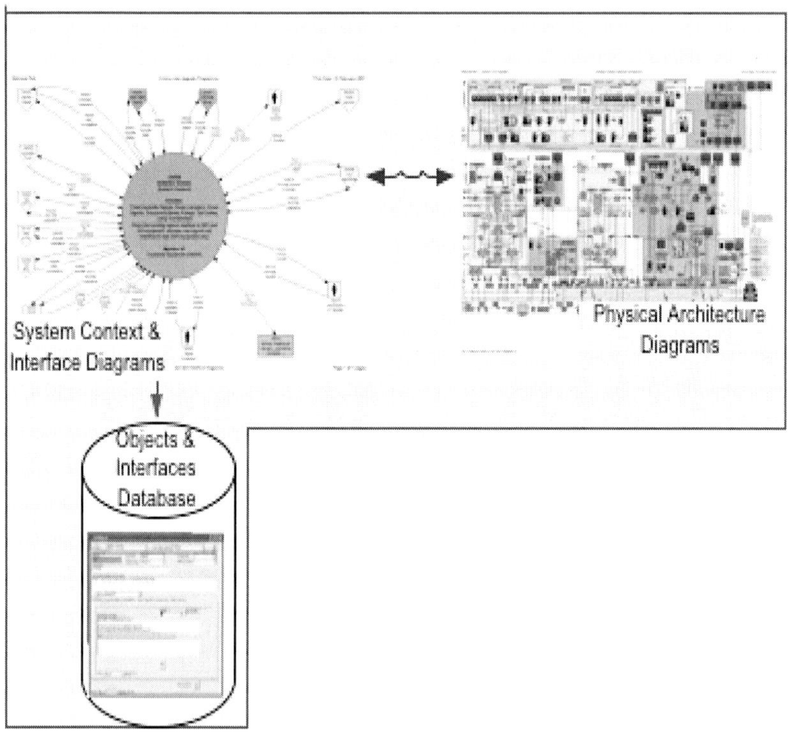


Figure 6-10: An example of System Definition

In line with the earlier definition of the system constituents' interfaces being categorised into Information, Energy and Action type interfaces, the system constituent symbol consists of a rectangle depicting the system constituent and three separate partitions of the rectangle, one for each interface type, as shown below on the Figure 6-11.

The agreed convention is that the interface type is defined on the source side of the interface, as it is the source system constituent that defines the type of the interface. Environment is treated as a special type of the constituent. Further formalisation of this methodology is introduced as part of the research, as follows.

Interfaces can be grouped into intentional and unintentional. Intentional interfaces are planned, designed interfaces, for example, communication links between two subsystems. This is why an intentional action or information type interface can only potentially be an indirect cause of the hazard or the consequence, and hence can only lead to the unintended or uncontrolled release of energy, when it fails to perform as desired.

Unintentional interfaces are not planned and appear either by a mistake in design, project implementation or as a result of a system failure. An interface that is not intentional is potentially a direct or indirect cause of a hazard or a direct or indirect cause of a consequence. It is only an unintentional interface of energy type that can be a direct cause of a safety consequence (in a sense of uncontrolled or unintended release of energy). An action or information type unintended interface can only be an indirect cause of the hazard or the consequence, and can only lead to the unintended or uncontrolled release of energy.

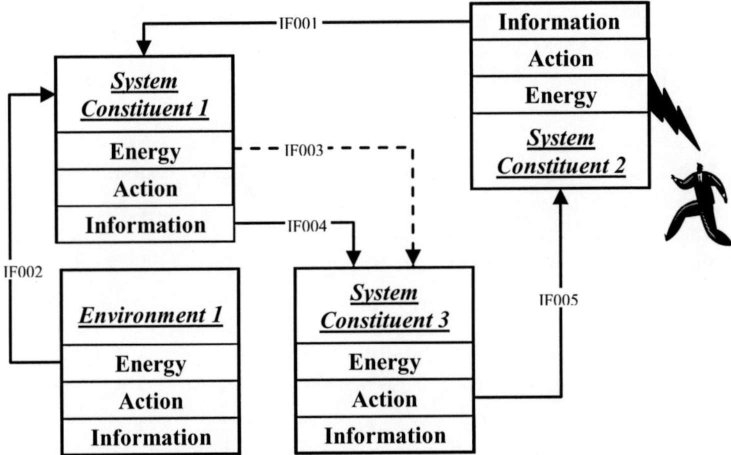


Figure 6-11: An example of the basic Process Model

Therefore, on the process models it is possible to distinguish between:

1. Intended interfaces (annotated by a full black line, for example comms link between the signalling system and on board computer);
2. Unintended interfaces that are indirect cause of the hazard or the consequence (annotated by black dashed line, for example Electromagnetic Interference) and

3. Unintended interface that is a direct cause of the consequence (annotated by a red lightning bolt).

All of the Process Model interfaces are described at the level of approximation adopted for development of process models. This description needs to include all changes to the interfaces to be implemented by the undertaking. Furthermore, it is suggested that the process model interfaces should be further "tagged" using a four dimensional classification system:

1. Primary classification: what is the function provided by the interface, or what is the purpose of the interface, reason for its existence;
2. Secondary classification: "being" of the interface, or a method of its execution;
3. Tertiary classification: temporal and operational nature of the interface, or when is the interface active, including all of the operational modes of the system;
4. Quaternary classification: physical environment of the interface, or what is the surrounding of the interface.

A set of generic classifications for a railway system have been developed and is detailed later in the book.

Since at the level of approximation of Process Models, some or all of the interfaces are conceptual or composite further cross-referenced to SCID interfaces, for detail and traceability is required. For example, communication between the train driver and a signaller on the process model would typically be presented with a single line interface, whilst in reality there are several different methods for this interface to occur, train radio, signal post telephone

It is of utmost importance to develop these models at the appropriate level to ensure that all relevant interfaces are included in the analysis, but also to make sure that the level of analysis is not too detailed (in order to avoid the loss of focus on the system analysis as opposed to detailed constituents analysis).

To ensure the completeness of the Process Models, it is proposed that a SCID is developed as a detailed system description model, as already discussed above, and that then clear boundaries between main system constituents, taking into account the engineering and operational aspects of the system as well as commercial/contractual boundaries related to the supply chain and the system constituents being supplied by different provider, are identified.

SCID and Process models are used in support of the information gathering process, and are inseparable as a system presentation tool.

Later in this book, the experience of a real world use of the process model technique will be discussed, and further formalisation will be proposed.

### 6.4.4 Initial Change Safety Analysis

The Initial Change Safety Analysis (ICSA) is done at two levels. Firstly, when the system definition is established, each of the boundary interfaces identified within the process models should be subjected to ICSA as already discussed earlier. This is essentially a structured review process, wherein interfaces are systematically reviewed by a group of experts against predefined criteria as explained earlier.

As part of the ICSA, in support of the future Change Safety Analysis, following information should be collected:

1. A description of the change to an interface;
2. What is the rationale behind the way in which the changes are applied;
3. What is the intended purpose of the new function;
4. What is the environment in which the new equipment will be installed;
5. What does the new equipment interface to;
6. What non-equipment changes, if any are being made at the same time (for instance, changes to the maintenance regime or operational rules);
7. Which aspects of the equipment and its application are novel and which have been proven in service;
8. Safety benefits of the change, if any;
9. ICSA category.

As a project is progressing, it is inevitable that the intended system definition will change, different subsystem constituents may not be fully compatible or new requirements may be identified, resulting in a different system configuration being brought into existence than the one that was planned. The SCID and the process models must be updated to reflect these changes, and each of these changes must be subjected to ICSA.

### 6.4.5 Change Safety Analysis

The Change Safety Analysis (CSA) aims to integrate a formal approach to system safety hazard management, quantification of system safety requirements, and system safety assurance. The Change Safety Analysis and Risk Assessment Process encapsulate the following:

1. Change Safety Analysis:
    a. Identification of Hazard, Causes, Consequences and Defence/Protection measures;
    b. Analysis of Causes;
    c. Analysis of Consequences including Loss Analysis;
    d. Analysis of Defence/Protection measures;
2. Risk Assessment;
3. Derivation of Safety Targets and Quantified Safety Requirements;
4. Change Safety Management.

***Identification of Hazards, Causes, Consequences and Defence/ Protection measures***

As argued earlier in the book, the hazard identification is the key phase of the safety analysis of any system. The knowledge gathered during this phase is a base for further analysis. The most important facet of the safety analysis is completeness. It is necessary to assure that the whole system has been analysed, and that all aspects of the systems behaviour have been scrutinised. To demonstrate completeness of the analysis, it is necessary to approach CSA systematically.

As previously discussed, the result of standard hazard identification is usually a linear collection of records that fit into different parts of the hazard space: hazards, causes, mitigations, consequences or just project/product issues not directly related to safety performance of the analysed system. Usually, regardless of the method used to support it, the quality of the outcome of the HAZID/HAZOP study depends only on the skills and experience of the person in charge of the study, and his/her ability to control and guide the participating experts. Most of the time, material gathered during these studies requires a significant amount of post processing and structuring.

With the aim to provide the structure to the HAZID/HAZOP studies and enable comprehensive understanding, the following approach is suggested:

1. Since the process models have already been produced for each core hazard, identification of hazards themselves has already being done

as part of conceptualisation. The first step of the CSA is to subject those interfaces that are classified as ICSA 1 or 2 to Change Safety Analysis (CSA) to identify related Consequences, Causes and Defence/Protection measures and, for those Defence/Protection measures that are not implemented yet, identify the Actions to implement them.

During the CSA it is useful to follow the guide words [Table 4-7] as it provides additional structure to the analysis. Also, it is necessary to consider all operational modes that the system may operate within (normal operation, emergency, degraded mode, failed mode, etc.) and within the different phases of its lifecycle (design, operation, maintenance commissioning/implementation, decommissioning, disposal).

Obviously if, as part of the CSA, a new core hazard is identified, the process needs to be reassessed, new process models produced for these core hazards and the process should recommence;

2. It is essential to keep the traceability from interfaces defined within the SCID to the interfaces of the Process Models and to the identified Hazards, Consequences, Causes and Defence/Protection measures. This will enable rigorous change control and assessment of:
    a. Impact of any changes to the original scope on safety analysis done thus far; and
    b. Impact of changes on the safety performance of the system.

***Causes***

Attributes of Causes are already defined earlier in the book; therefore, here we will focus on methodologies for analysis, assessment, and identification of the relationships and possible combinations between the causes.

It is proposed that a distinction between three areas of the analysis is made:

1. Characterising the Causes.

   The aim of this activity is to define the Cause in terms of its description, origin, location and period/time when it is active.

   This is done as part of CSA. A description of the cause must be clear, and must provide sufficiently detail to allow further analysis. Since the CSA is carried out by a systematic analysis of the process models' interfaces, identification of the origin and spatial and temporal characteristics of the cause is achieved through mapping these characteristics from the originating interfaces to the related causes;

2. Assessing the measure of the contributions of the Causes to a Hazard and subsequently to any consequences arising from the hazard.
   Detailed discussion about this analysis is provided later in section on Risk Assessment;
3. Correlating/Associating the Cause with other Causes, Hazards and Defence/Protection measures.
   It is often the case, that failure of some interface (failure of a function provided by the interface) may lead to more than one hazard or it is necessary for some other interface to fail in order for the hazard to occur; i.e. a combined failure could lead to a hazard. Also some of the causes will actually lead to failures of the Defence/Protection measures themselves.
   It is necessary to identify all of these relationships in order to comprehensively understand the effect of the interface failures or in other words, Causes, on the overall system.

Analyses of the causes could be supported by several different methods, for example:

1. FMEA/FMECA (see earlier discussion) is usually performed at the detailed design level, whereby an analysis of the design is undertaken to identify potential failure modes. This approach can be utilised for analysis at the system level;
2. Reliability Block Diagrams and Network Models (see earlier discussion), again are methodologies usually utilised for detailed analysis at the design level in order to identify common mode failure modes and enhance system redundancy.

***Consequences and Loss***

An understanding of the potential consequences of a hazard is a prerequisite to an appreciation of the risk and, therefore, to management of the consequence.

Attributes of consequences have already been discussed, as well as different approaches to the estimation or calculation of probability, or the frequency of consequence occurrence and loss analysis.

Therefore, here the author will only discuss several aspects of the analysis of consequences, and the attributed losses:

1. It is necessary to define the location of consequences and their temporal nature as part of the initial identification. For example, if a consequence happens only during out of normal service hours, the loss associated with that consequence could be much less than if it happens during peak hours when exposure to the consequence is much higher;
2. As part of the initial analysis, relationships between the consequences should be identified. It is possible that the consequence identified initially is actually a secondary effect of some other initiating consequence, or that it only materialises in parallel with another consequence.

It is also important to identify the nature of subsequent loss in terms of safety, commercial and environmental loss and the exposed party (service users, service operators, neighbours).

### *Defence/Protection Measures*

Mitigations, Barriers and Containment Measures and their attributes are already defined earlier, so here we will focus on only those aspects of identification and analysis that have not already been discussed earlier.

Assessing the measure of effectiveness of Defence/Protection Measures is a vital enabler towards a conclusive ALARP argument. An assessment of effectiveness of Defence/Protection Measures depends on a comprehensive understanding of:

1. "Being" of these, in terms of their origin (are these measures a physical thing, an alarm for example, a planned human intervention such as scheduled maintenance or a circumstantial event);
2. The temporal and spatial nature of the Defence/Protection Measures, or when and where these measures are active and effective;
3. The potential relationship between different Defence/Protection Measures, for example it is possible that Defence/Protection Measures are effectively working "against" one hazard but may actually contribute to another (increase number of maintenance checks for example may reduce a number of equipment failures but may increase exposure of staff to occupational hazards);
4. Reasons, or causes for, any failures of the Defence/Protection Measures; as these may be also related to causes of a Hazard.

### *Risk Assessment - Quantitative versus Qualitative*

Both of these approaches have been discussed in detail earlier, and it is not the intention to detail here the techniques and different methodologies for risk assessment.

Both approaches are equally legitimate and informative, and each suffers from some common and specific limitations.

The main shortcomings of quantified risk assessment (QRA) are cost (often the cost of this activity cannot be easily justified, particularly for simple undertakings) and lack of reliable data (particularly if the undertaking is about a novel and/or complex change). On the other hand, qualitative risk assessment often results in unsubstantiated arguments based around informal estimates of consequence occurrence and associated loss. In addition to that, some problems are very difficult, if not impossible, to model; for example, system failures related to human action and software performance, without experiential data being available.

As already discussed earlier, it is often possible to develop a substantiated quantified risk assessment model of an existing system, and in these circumstances, most of the time it is possible to obtain valid data to support the modelling of this existing system. It is valuable, as it is rarely the case that the whole system is introduced as a completely new novel system. Most often, the undertakings are about the existing systems (for example railway systems) being upgraded or improved through either a number of small changes in time or a substantial large scale upgrade programme. In these cases it is possible to develop the QRA model representing the safety performance of the system before any changes to the system are made. This approach permits an initial model to be populated with valid data and gauged against the validated historical data. This model can then be used as a base for the development of a model representing the effect of the changes to the system and the impact of these changes to the safety performance of the system. This approach minimises the amount of new data to be identified/derived and makes it possible to validate the model's predictions against a reliable reference point.

However, regardless of how much validation and verification of data is done, in practice the use of the absolute value predictions of common QRA models (the most often technique used within the railway industry is the fault tree) in safety reasoning is very arduous to assure and justify.

Most often, in practice, use of the QRA predictions is limited to:

1. Comparison of safety performance of a future system against the safety performance of an existing system. This is always done in relative terms, the outcomes of the two models, are compared and

the difference between two outputs is taken as the result of the analysis. This is then used in support of an argument that the safety performance of a future system is in relative terms better or at least the same as the safety performance of an existing system. In the case that the analysis shows that the safety performance of a future system is worse than that of an existing system, further analysis and changes to the future system must be undertaken until the results are favourable. For more detail on this please see the next section of the book;

2. QRA models are often used to set/derive initial targets for a developing system. Later on, once the constituents of the system have been developed and tested, real data can be fed back into the model to confirm that the overall safety target has been achieved.

Where a qualitative approach is applied, it is suggested that instead of an absolute estimate of the risk, a comparative approach is taken, as described earlier in section "Complex Railway Project Safety Management – Manchester South Capacity Improvement Project". It is along the same line of reasoning as that adopted for the QRA modelling; if a reference point is defined and used for the measurement of risk, the claim for assurance of the assessment should be built around the comparison with a known risk profile. Using this methodology greatly enhances the quality of expert estimates; qualitative assessment is almost always based on knowledge of domain experts and inference.

In its nature the Inquiry System suggested here is a Lockean Inquiry System, the problem domain is an empirical one, building arguments into a fact net, based on a set of empirical judgments elicited from domain experts.

### *Risk Assessment - Apportionment, Optimisation, Options and Impact analysis, Amalgamation of Quantitative and Qualitative Analysis and argumentation for selection of options for implementation*

A systematic study, of each hazard, regarding the causal and consequential effects of failures, shall be carried out. In order to identify all the applicable defence/protection options, the best available expertise shall be employed. The results of the apportionment, and the importance analysis, are used as guidance for the concentration of effort and the prioritisation of options analysis.

To enable a rigorous option suitability assessment, the following information must be provided for each identified option:

1. Cost of implementation;

2. Timescale for implementation;
3. Need for additional training;
4. Impact of implemented option on a selected consequence;
5. Impact of the selected option on the overall safety performance forecast by the model.

This information will provide a basis for the decision-making process and support the ALARP argument.

As explained, both approaches to risk assessment, Quantitative and Qualitative, have some limitations, mainly caused by the difficulty in obtaining reliable data to feed into the reasoning model. To overcome this problem it is proposed that both approaches are used, to complement each other, in support of the safety risk assessment and justification.

In general, the construction of the safety argument should be carried out around two lines of reasoning (Health and Safety Executive, 2009):

Demonstrating that the safety performance of a new or changed system is at least as good as the safety performance of a similar or previous system (before any changes are done to it).

Ethically and legally, we expect the new or enhanced system to deliver at least as good a safety performance as the previous one. With the rapid advance of technology and science it would be impossible to justify increasing the societal exposure to risk; the expectation is to improve safety performance with time.

Using the QRA methodology in support of this line of reasoning is beneficial, and compatible with the approach detailed above.

Demonstrating that the owner and/or the beneficiary of the undertaking has reduced the safety risk emerging from the system to As Low As Reasonably Practicable (ALARP).

Ethically and legally, it is expected that the risk of any undertaking is mitigated/controlled to an acceptable level. In the UK in particular, the acceptable level of risk is defined as ALARP. This requirement calls for the identification of the controls of hazards on an individual basis (for each hazard) including consideration of all possible controls. It is only when there is nothing else practicable that can be done to reduce the risk level further, that the safety risk emerging from the system can be considered as ALARP.

To support the argument that the safety risk emerging from the system is ALARP, it is also necessary to demonstrate the practicability argument by carrying out a cost-benefit analysis to assess and compare the costs of any proposed changes to the system against the reduction in risk levels inclusive of the acceptance/rejection criteria.

It is proposed here that the analysis should be done at two levels:

- For specific highly critical hazards (ones that have major contributions to safety risk) the QRA approach is recommended, but only if it is possible to develop QRA models representing the system before changes or a similar already existing system so that data and the model predictions can be validated and verified. The QRA should then be used in support of the options identification and analysis and the ALARP argumentation.
- At the system level, for each hazard, a qualitative/comparative methodology suggested earlier (please see section "Complex Railway Project Safety Management – Manchester South Capacity Improvement Project") should be applied.

This methodology allows for a systematic and transparent assessment of all hazards and related causes, and defence/protection measures. However, in order to enable the process it is necessary to structure the hazard log in such a way as to make it possible to comprehend relationships between different elements of the "hazard universe", and to ascertain sufficient information to support domain experts in their review and estimation of risk.

### *Derivation and Management of Safety Requirements*

In recent times with the advance of systems engineering, we find that that management of interfaces between system constituents, suppliers, contracts, and internal divisions delivering parts of the system, etc. is executed through the management of system requirements. The system requirements should say what the system must do, not how it must be done, and should be set at the level of interfaces across boundaries within the system, where the boundaries are defined as boundaries between major system constituents, and system and environment themselves. The boundaries may be dictated by contractual arrangements for the delivery of the system constituents, internal divisions into delivery groups on the project, etc. Regardless of the reasons behind the structure of the project delivery organisation, these influences must be taken into account when system boundaries are defined within the system architecture and/or process models as described earlier.

The Change Safety Analysis Process, Safety Requirements should be derived from identified defence/protection measures. However, not all of the defence/protection measures are capable of being translated into requirements. Some defence/protection measures are simply statements of fact, and although valid, these are not worthy of being transformed into requirements.

For example, the Victoria line of London underground is located wholly underground in a deep tunnel. It never rains on the Victoria line and this is a good mitigation against corrosion, but it is not a requirement. Furthermore, in line with the approach taken regarding the Change Safety Analysis process, if the change is conventional (like for like replacement or introduction of already proven and used technology), there is no need to define detailed requirements as such, a reference to existing standards and good practice should suffice. On the other hand, the introduction of a new signalling system demands the systematic identification of all related requirements, including safety requirements, and methodical management to validated and verified implementation.

Some of the safety requirements are not quantifiable as such. For example, a safety requirement that "all train drivers must be trained to drive the train in all operational conditions" is a binary statement, it is either done or not.

If we define a set Mh of all defence/protection measures (mitigations "m", barriers "b" and containment measures "c") of a hazard h to be Mh = {m, b, c}, then set of all safety requirements Rh, associated with the hazard h, is a subset of set M; Mh $\subseteq$ Rh.

Furthermore, a set of quantified safety requirements Qh, associated with the hazard h, is a subset of set Rh; Mh $\subseteq$ Rh $\subseteq$ Qh.

Selected defence/protection measureS should be stated in the form of safety requirements on performance, function, procedure, environment, competence and other reduction methods. The safety requirements must then be apportioned to those projects and suppliers that are responsible for delivery of the equipment, people and procedures that embody the causal factors identified in the analysis.

The projects and suppliers, to which these safety requirements have been apportioned, are responsible for supplying the evidence that safety requirements have been met. The System Safety Case is populated with the system safety requirements as apportioned to the projects and suppliers and evidence coming from the projects and suppliers.

The safety requirements are associated with the defence/protection measures they have been derived from and should be managed through the hazard log, because implementation of a requirement enables closure of a hazard log entry.

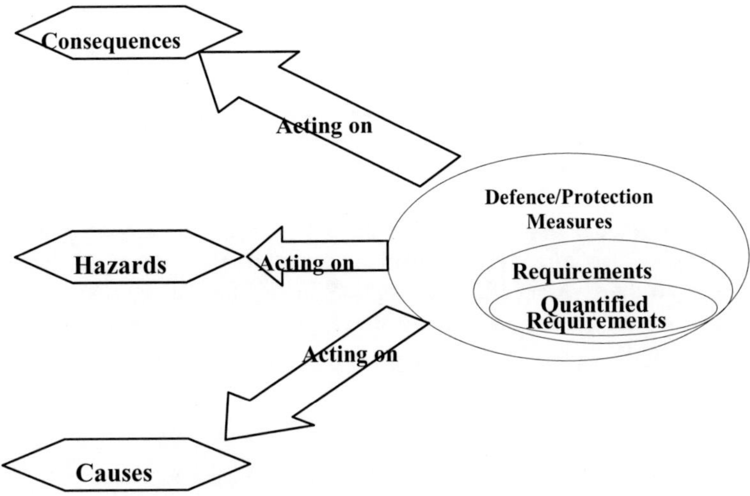


Figure 6-12: Defence/Protection measures, Requirements & Quantified Requirements

### 6.4.6 Assertion of Completeness of System Analysis

An essential part of any safety argument is an argument about the completeness of the safety analysis. In making any logical claim, the most difficult part of the argument is proof of completeness of the analysis, and an understanding of the uncertainties involved in the decision making process. The following are the areas of interest:

1. Completeness of the system understanding:
   a. understanding of the existing system (system to be changed);
   b. understanding of the changes to the system to be introduced;
2. Completeness and correctness of Change Safety Analysis:
   a. completeness of hazard identification and identification of protection measures and safety requirements;
   b. correctness of risk analysis and assessment and certainty of data used in support of risk assessment;
3. Completeness of the safety arguments/facts net:
   a. completeness of totality of lines of reasoning explored;
   b. completeness and certainty of evidence provided in support of arguments.

### *Completeness of the system understanding*

Completeness of the system understanding consists of an understanding of the existing system, the system before changes and the changes to be introduced to the system and the effect of these changes on the system properties.

The process models have been used on both projects to support this understanding, as has already been explained earlier. Confidence in the completeness of the system understanding is achieved through the rigorous application of the process models', their production, review and configuration control.

### *Completeness of Change Safety Analysis*

There are two slightly different types of completeness and certainty in relation to Change Safety Analysis.

The first type is about the comprehensiveness of an analysis of the conceptualised system (process models) that is aimed at the identification of hazards, protection measures and requirements.

The second type is about the correctness of the risk assessment and analysis of the identified hazards, protection measures, requirements and the correctness of information used in the analysis. It is particularly relevant to QRA and to options and impact analysis.

The assertion of comprehensiveness is achieved through adherence to a structured and rigorous process of hazard identification; preliminary analysis (Initial Change Safety Analysis and Change Safety Analysis) is done on the process models of a system as base-lined and agreed by the project. The analysis must include all of the boundary interfaces identified within the system architecture and mapped to the process models.

Once the system architecture and process models are base-lined, any changes to the system should be controlled. The change control mechanism is usually known as the Engineering Change Request, and is in a form of an official request for a change to a base-lined design. Usually this process is not formalised in the sense of changes being referenced against interfaces, rather, the change is described and the impact of change is assessed by competent experts.

However, to comprehensively support ICSA and CSA processes it is necessary to integrate configuration control of the system architectures with the ECR process. It is suggested that each change request is referenced to the interface impacted by the change, and submitted to the custodian of the system architecture for impact assessment. Any change to the system should be approved by the custodian of the system architecture and the affected system architecture/process models updated.

In order to complete the impact assessment, the effect should be assessed on the safety performance of the system and/or effect of changes to safety analysis already completed. For each change, affected interfaces should be identified, and then subjected to ICSA and if necessary CSA. The output of this analysis feeds into a hazard log providing the base for further analysis and supporting construction of safety arguments; i.e. the amount of rework to the ICSA and CSA must be defined and then completed.

The assertion of correctness is achieved through the adherence to the rigorous and transparent process for development of the QRA logic, data collection and review.

However, in this case it is also possible to gain further confidence with regards to the data used for QRA. The use of sensitivity analyses to identify the most critical data items, focused data searching and checking, use of multiple, independent data sources, comparison of model prediction with real life data (base-lining) are all techniques that should be used in support of building confidence in the modelling results.

### 6.4.7 Hazard Log, tool for Change Safety Management, Configuration Control and Reporting

The hazard log is a central repository of the knowledge gathered during the analysis and the tool supporting the management of activities related to safety analysis and control.

All the information gathered through the analysis should be captured in the hazard log. The basic information to be contained in the hazard log is defined in the railway industry guidance, the Yellow book (RSSB, 2007):

*Journal* should describe all amendments to the Hazard Log, in order to provide a historical record of its compilation and provide traceability.

*Directory*, sometimes known as the Safety Records Log, should give an up-to-date reference to every safety document produced and used by the Project. It may be convenient to keep the Directory separate from the rest of the Hazard Log, or even to integrate it with a project document management system.

*Hazard Data* should record every identified hazard. For each hazard, the information listed later in this section should be recorded as soon as it becomes available. Data collected during Hazard Analysis and Risk Assessment should be transcribed to the Hazard Log when the reports have

been endorsed. If the hazard is not closed or cancelled, then the name of a person or company who is responsible for progressing it towards closure should be captured as well;
*Incident Data* should be used to record all safety related incidents that have occurred during the life of the system or equipment. It should identify the sequence of events linking each accident and the hazards that caused it.
*Accident Data* should be used to record every identified possible accident. It should identify possible sequences of events linking an identified accident with the hazards that may cause it.

It is easy to see that all of the information listed above has been already identified as part of the overview of the hazard attributes earlier in the book. The only outstanding part is the reference to Directory and Documents, and this is not directly related to identification, analysis and management of individual hazards. The author would recommend that generic documentation related to plans and processes for safety management on the project should not only be included in the hazard log directory, but should be integrated with the project document management system.

Since all of the information gathered during the Change Safety Analysis is contained in the hazard log, it seems logical to utilise the hazard log to actively support further analysis, not only as a repository of results of the analysis. If the hazard log is structured in such a way as to present the data in a form of a Meta-model of the "hazard universe", it could then be used in support of further analysis, for example in support of qualitative risk assessment, review for completeness, or construction of safety arguments.

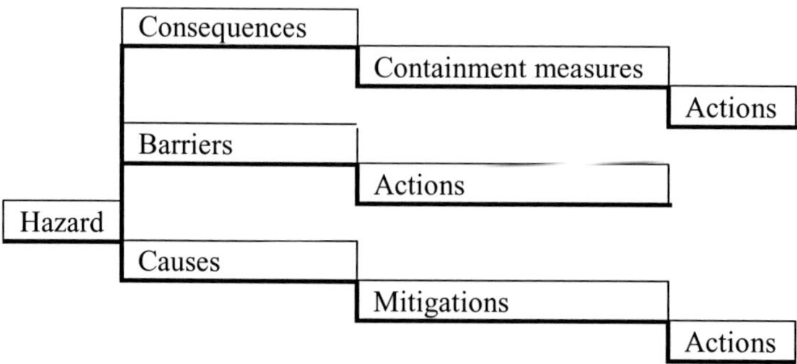


Figure 6-13: Hazard Log Hierarchy

With this in mind, a Hazard Log database should be developed. The complexity of the Hazard Log database is due to the many – to – many relationships that occur between the different elements of the hazard log; for example, one consequence may apply to many hazards, and one hazard may have many consequences. This cross-fertilisation between the elements lends itself to the capability of cross referencing to other elements by selection. The figure below shows the relationship between the data items (tables) in the Hazard Log:

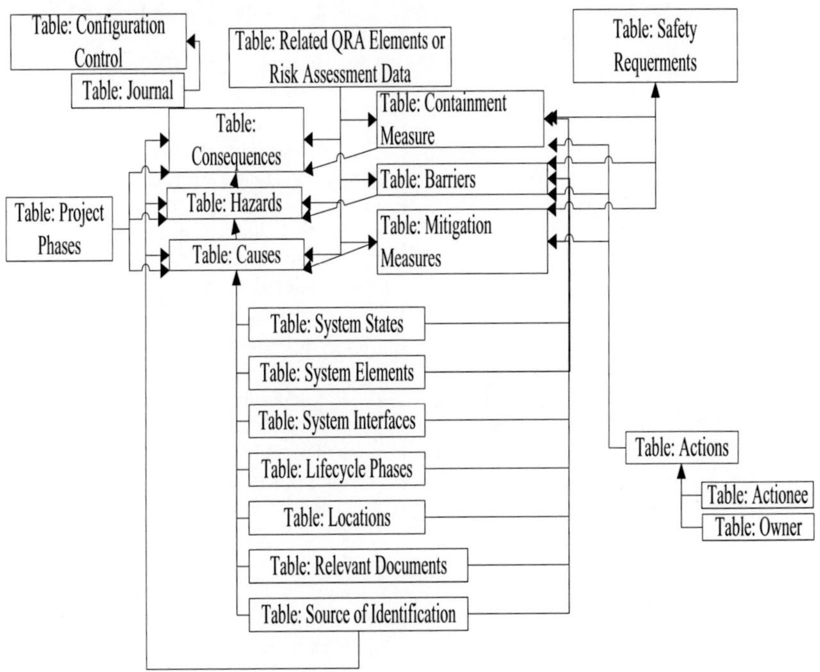


Figure 6-14: Hazard Log relationships

Note: all tables are in one-to-many relationships, in the direction as shown in figure 6-14 above (for example, one cause can relate to many hazards) except for the relationships between:

- Table: Consequences and Table: Hazards, where each consequence can belong to only one hazard and
- Table: Configuration control and Table: Journal, where each journal entry can be related to only one configuration control entry.

The following table describes the various elements that make up the Hazard Log and its relationship to other elements:

| Element Name | Attributes | Attributes-Relationship to elements |
|---|---|---|
| Hazards | 1. Unique Identifier;<br>2. Hazard Name;<br>3. Hazard Description;<br>4. Comments;<br>5. Disposition Statement. | Many – to – many relationships with:<br>1. Related Causes;<br>2. Related Barriers;<br>3. Related QRA Model Elements (if QRA is used);<br>4. Pertinent Lifecycle Phases;<br>5. Pertinent Project Phases;<br>6. Source of Identification;<br>7. Relevant Documents;<br>One to one relationship with:<br>1. Related Consequences;<br>2. Hazard status (Open; Conditionally Closed; Cancelled; Closed; Transferred);<br>3. Pertinent Location. |
| Causes | 1. Unique identifier;<br>2. Cause Name;<br>3. Cause Description;<br>4. Measure of a contribution of the cause to the Hazard and related Consequences;<br>5. Comments. | Many – to – many relationships with:<br>1. Related Hazards;<br>2. Related Mitigations;<br>3. Related QRA Model Elements (if QRA is used) or Risk Assessment Data;<br>4. Related System States;<br>5. Related System Elements;<br>6. Related System Interfaces;<br>7. Pertinent Lifecycle Phases;<br>8. Pertinent Project Phases;<br>9. Source of Identification;<br>10. Relevant documents;<br>One to one relationship with:<br>1. Status (Open; Conditionally Closed; Cancelled; Closed; Transferred)<br>2. Pertinent Location. |
| Consequences | 1. Unique identifier;<br>2. Consequence Name;<br>3. Consequence Description;<br>4. Loss associated with the consequence; | Many – to – many relationships with:<br>1. Related Containment Measures;<br>2. Related QRA Model Elements (if QRA is used);<br>3. Related System States; |

| Element Name | Attributes | Attributes-Relationship to elements |
|---|---|---|
| | 5. Frequency or probability of consequence occurrence<br>6. Comments;<br>7. Initial Risk;<br>8. Residual Risk. | 4. Related System Elements;<br>5. Related System Interfaces;<br>6. Pertinent Lifecycle Phases;<br>7. Pertinent Project Phases;<br>8. Source of Identification;<br>9. Relevant documents;<br>One to one relationship with:<br>1. Related Hazards;<br>2. Exposed party;<br>3. Pertinent Location. |
| Mitigation Measure | 1. Unique identifier;<br>2. Mitigation Name;<br>3. Mitigation Description;<br>4. A measure of reduction of the detrimental effect of the cause;<br>5. Comments. | Many – to – many relationships with:<br>1. Related Causes;<br>2. Related Safety Requirements;<br>3. Actions for implementation of mitigation;<br>4. Related QRA Model Elements (if QRA is used);<br>5. Related System States;<br>6. Related System Elements;<br>7. Related System Interfaces;<br>8. Pertinent Lifecycle Phases;<br>9. Pertinent Project Phases;<br>10. Source of Identification;<br>11. Relevant documents;<br>One to one relationship with:<br>1. Pertinent Location;<br>2. Status (Implemented; Not Implemented) |
| Barrier | 1. Unique identifier;<br>2. Barrier Name;<br>3. Barrier Description;<br>4. A measure of reduction of the detrimental effect of hazard<br>5. Comments. | Many – to – many relationships with:<br>1. Hazards;<br>2. Safety Requirements;<br>3. Actions;<br>4. Related QRA Model Elements (if QRA is used);<br>5. Related System States;<br>6. Related System Elements;<br>7. Related System Interfaces;<br>8. Pertinent Lifecycle Phases;<br>9. Pertinent Project Phases;<br>10. Source of Identification;<br>11. Relevant documents;<br>One to one relationship with: |

| Element Name | Attributes | Attributes-Relationship to elements |
|---|---|---|
| | | 1. Status (Implemented; Not Implemented);<br>2. Pertinent Location |
| Containment Measure | 1. Unique identifier;<br>2. Mitigation Name;<br>3. Mitigation Description;<br>4. A measure of reduction of the loss;<br>5. Comments. | Many – to – many relationships with:<br>1. Consequences;<br>2. Actions;<br>3. Related QRA Model Elements (if QRA is used);<br>4. Related System States;<br>5. Related System Elements;<br>6. Related System Interfaces;<br>7. Pertinent Lifecycle Phases;<br>8. Pertinent Project Phases;<br>9. Source of Identification;<br>10. Relevant documents;<br>One to one relationship with:<br>1. Status (Implemented; Not Implemented);<br>2. Pertinent Location. |
| Actions | 1. Unique identifier;<br>2. Action Name;<br>3. Action Description;<br>4. Comments;<br>5. Action Response<br>6. Information Supporting Closure;<br>7. Implementation Deadline;<br>8. Allocation Date;<br>9. Completion Date. | Many – to – many relationships with:<br>1. Related Mitigation;<br>2. Related Barrier;<br>3. Related Containment Measure;<br>4. Source of Identification;<br>5. Relevant documents.<br>One to one relationship with:<br>1. Action Implementation Managers;<br>2. Status. |

**Table 6-1: Hazard Log Elements and Relationships**

| Element Name | Attributes | Attributes-Relationship to elements |
|---|---|---|
| Configuration Control | 1. System hazard log version number;<br>2. Version details;<br>3. Comments;<br>4. Issued by | One to Many relationship with:<br>1. Related journal entries |

| Element Name | Attributes | Attributes-Relationship to elements |
|---|---|---|
| Journal | 1. Journal Entry Reference;<br>2. Entry created by;<br>3. Date of journal entry;<br>4. Comments. | |

**Table 6-2: Configuration control and Journal data**

If the Hazard Log is structured in such a way, and contains all of the information discussed above, it could be used in support of the following safety analysis and management activities:

1. Structuring and grouping of the identified causes, mitigations, hazards, barriers, consequences and containment measures into Core Hazards' meta models supporting further analysis;
2. Management of safety activities through management of actions;
3. Configuration management of safety information and audit trail;
4. Reporting against the status of the hazards analysis, management of safety activities and achieved levels of safety.

### 6.4.8 Continual Appraisal and Knowledge Update

Through the life cycle of the project/undertaking, it is necessary to keep the information required to secure the safe delivery of the undertaking up to date. This should be implemented through the regular review of changes to the original system design, for example, the review of ECRs and the impact of these changes on the analyses completed so far, and the review of data collected by the system performance monitoring processes such as Failure Reporting And Corrective Action System (FRACAS).

During the lifecycle of the project, in particular in the early stages of the lifecycle, and in order to initiate analysis, it is necessary to make some assumptions, for example, about the future performance of the system, or about particular interface behaviour. Furthermore, during the analyses, often it is necessary to identify dependences that need to be acted upon prior to the operation of the system, and caveats or restrictions setting the limits, defining the boundaries of the operational envelope of the system. These need to be regularly reviewed and updated. A change of assumption may initiate additional analysis, or a review of analysis already completed, to ensure that the results are still valid. A change of dependency may cause a change of schedule, a new chain of actions to be implemented and finally, a change in restrictions to be implemented. These need to be captured as it is the restrictions that have a significant contribution in defining the safe operational envelope of the system being implemented or changed. For the purpose of the management and control of these, it would be beneficial to establish databases to collate and control Change Requests and Assumptions, Dependences and Caveats (Restrictions).

## 6.5 Chapter Conclusions

In this chapter, the author presented and discussed the new systems based framework for the analysis and management of safety risks that the author developed against the requirements identified earlier in Chapter 5.

In support of the framework, the author analysed the hazard as a system concept and developed existing methodologies further, identified and defined the attributes, and their hierarchical structure, of a Hazard as a system. The author also analysed the concept of the 'safety case' as the inquiry model identifying a number of constraints and requirements that a safety case must satisfy. Subsequent to the analysis of these, the author proposed the general high level approach to the structure of the safety case as a fact net, reasoning model.

In previous chapters, the author presented the results of the analysis of the existing system safety and system engineering processes and guidance, and some of the early development results, and identified requirements for a new framework. The author outlined the vision of safety engineering as an integrated part of the system engineering process, and the assurance of delivering the needed functionality of the four key emerging properties of any system: Safety, RAM, Operability and Performance.

Following on from that, the author developed and outlined the integrated process for the analysis and management of engineering safety. The process consists of 7 steps, most of which are not newly identified as needed, but have not been formalised either, nor used together within an integrated framework.

# CHAPTER SEVEN

# APPLICATION

## 7.1 Chapter introduction

The aim of this chapter is to present the results of a real life application of the safety process discussed in the preceding chapters. The scope includes two major London Underground upgrade projects; the Victoria Line Upgrade Programme and the Subsurface Lines Upgrade Programme.

The author, together with his team, has successfully applied the process depicted in this book to the former, and is currently applying it to the latter.

As part of the application of the process and methodologies, developed earlier, the need for further enhancements was identified and is also discussed in this section.

Figure 7-1 below outlines the structure of this section.

## 7.2 Background

The Victoria Line Upgrade Programme (VLUP) is a multi-disciplinary project. The principal works included within the programme was the upgrade of rolling stock, signalling and signalling control. The project was successfully completed in 2013. The solution involves:

1. Delivery of a fleet of 47 new trains, to be driven under Automatic Train Operation (ATO), but with the capability to be driven manually at full speed under Automatic Train Protection (ATP);
2. Delivery of a new signalling system, including:
    a. Train borne equipment,
    b. Signal equipment room equipment,
    c. Control Centre equipment, and
    d. Trackside equipment.
3. Delivery of a new signalling and communications control systems, located within a new Service Control Centre (SCC);
4. Removal and disposal of existing Victoria Line trains and signalling;

5. Provision of train cab and signalling simulators; and
6. Training in system operation and maintenance, and the provision of maintenance manuals and full technical documentation.

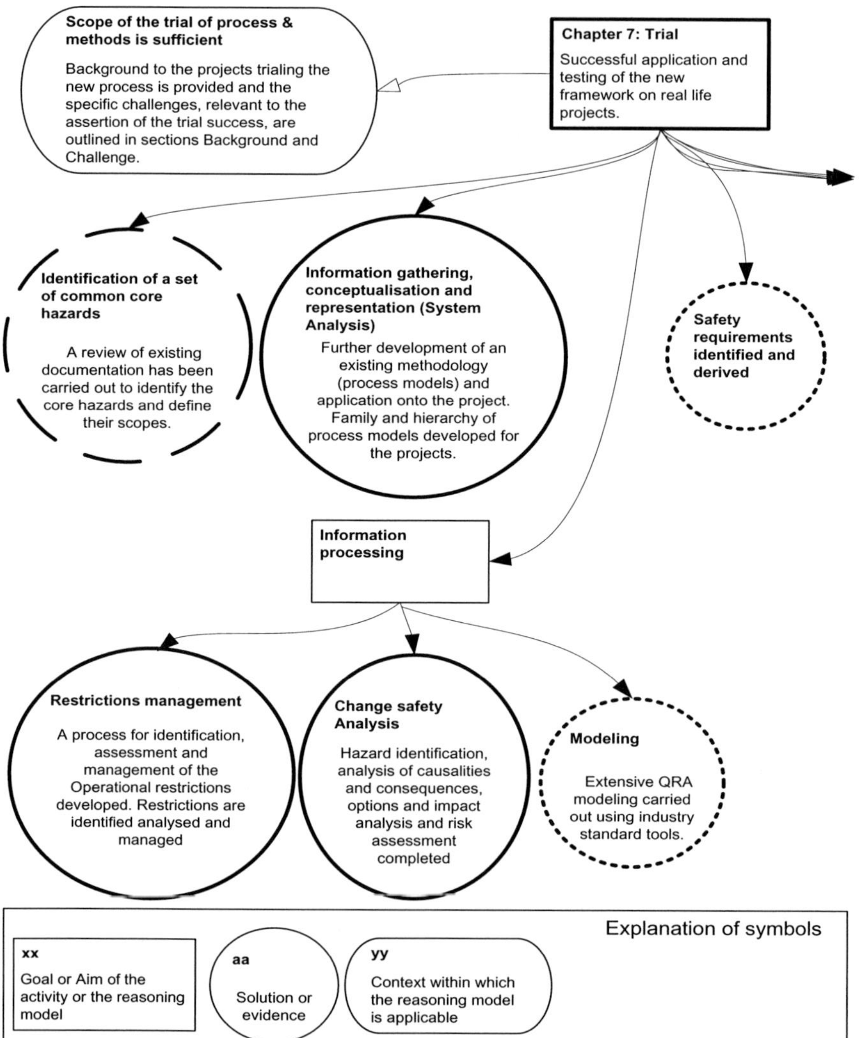

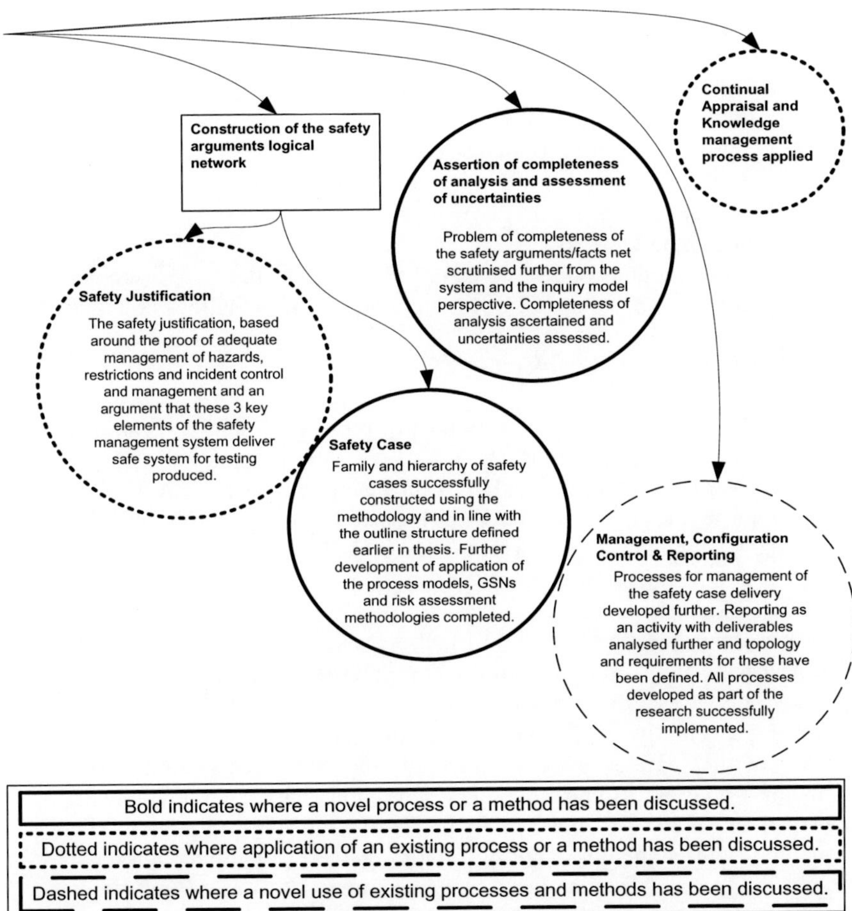


Figure 7-1: Structure of Chapter 7 and argument in support of the claim that the newly developed process has been successfully trialled and any shortcomings in the initially developed process developed further

These works were supported by a number of infrastructure enabling works and support activities. These included:

1. Provision of new Signalling Equipment Rooms (SERs) and cable routes;
2. Provision of new signalling power supplies with battery backup (Uninterruptible Power Supplies);

3. Track upgrade works (through the Metronet Rail Bakerloo, Central and Victoria Line (MRBCV) Track Renewals Programme);
4. Substation enhancements (through the Power Private Finance Initiative (PFI) with EDF);
5. DC power distribution enhancements by the VLUP;
6. Transmission links and train radio enhancement (managed through the Connect PFI);
7. Depot upgrade works;
8. Station upgrade and enhancement - including Customer Information Systems (CIS) (through the MRBCV Stations & Civils Programme);
9. EMC and gauging surveys.

The VLUP was delivered by a number of projects (around 25) and in a number of major configurations.

To deliver the VLUP, the VLU Project Team within the Victoria Line Programme was established. The VLU Project Team was responsible for managing all contracts entered into, to deliver the VLUP, for integration of the various subsystems that make up the upgrade and for various other cross-discipline functions.

It was the responsibility of the VLU Project Team to deliver the overall Engineering Safety and Assurance Case (ESAC), comprising of the System Safety Case, Railway Operational Readiness Report, System RAM Status Report and System Performance Status Report.

Bombardier Transportation supplied the trains, signalling and signalling control. Bombardier Transportation has a contract with Westinghouse Rail Systems Limited (WRSL) for the delivery of the signalling and signalling control, while delivery of trains was the responsibility of Bombardier Mainline and Metros. Bombardier Transportation was responsible for the delivery of these systems and their integration for the VLU.

Subprojects of the VLU Project carried out the delivery of the Victoria Line Service Control Centre (SCC) building and systems integration, and other enabling works such as track upgrades and SER construction.

Two Private Finance Initiatives (PFI), being managed through separate contractual arrangements by LU, also enable the VLUP:

1. The Power PFI contract with EDF Energy Powerlink Limited - responsible for power supply upgrades by way of substation enhancement;
2. The Connect PFI contract with CityLink Telecommunications Limited - responsible for radio and transmission upgrades.

There was no direct contractual relationship between VLUP and the two PFI suppliers.

Requirements from LU were placed on VLUP through the Public Private Partnership (PPP) contract. Any changes to these requirements needed to be negotiated with the client organisation within LU. LU was responsible for providing safety analyses and assurance evidence from the PFI suppliers to the VLUP to support the VLUP System Safety Case, as required by the VLUP System – wide Work Package Plan.

The VLUP also interfaced with other LU/Metronet Rail/Tube Lines supply chain partners responsible for other London Underground Upgrade programmes:

1. Balfour Beatty - responsible for ongoing track renewals;
2. Stations Alliance - responsible for refurbishment and modernisation of Metronet stations and civil infrastructure.
3. Station upgrades, carried out by Tube Lines.

All of the subsystem suppliers to the VLUP were responsible for safety management within their individual scope.

Operation of the line was the responsibility of LU (apart from engineering hours when it was responsibility of VLUP).

Maintenance of the line was the responsibility of the LU Maintenance Organisation. However, initially, WRSL would maintain the signalling part of the system with a planned hand over to the LU Maintenance Organisation at later stages in the project.

The Subsurface Lines Upgrade Programme (SUP) consists of many projects that together deliver the capability improvement requirements and PPP contractual obligations to achieve improved Subsurface Lines (SSL) railway performance.

The following major programme of works is to be undertaken for the upgrade of the SSL railway:

1. Replacing the ageing rolling stock offering improved capacity and performance;
2. Replacing the obsolete signalling control system with one offering greater operational management capability and improved headways;
3. Providing a new Service Control Centre (SCC) for LU controllers, ergonomically integrating a number of systems and their human factor requirements including signalling control, train communications, Public Address, tunnel telephone and auto telephones;

4. Providing the necessary power upgrades including the upgrade of low voltage AC power required for SUP and station upgrades, and to support LU with the DC traction power upgrade;
5. Necessary infrastructure enabling works such as track, station and depot upgrades including the provision of new signal track equipment;
6. Providing rolling stock (cab) and signalling control system simulators;
7. Training in the operation and maintenance of the new or altered assets and systems provided, as well as maintenance manuals and full technical documentation.

Each SUP Upgrade Step is delivered through a staged process of translating the system requirements and contractual obligations into infrastructure changes. The final stage, and hence the upgrade of the complete SSL railway, is due to be delivered by March 2018.

The principal objective of the staged process of commissioning the SSL railway is such that:

1. Safety risk is kept As Low as Reasonably Practicable (ALARP);
2. That the risks from the SSL upgrade programme are understood and managed to acceptable levels as the new systems are rolled out;
3. As the new systems are rolled out, revenue service is maintained at existing levels, with service improvements where possible;
4. The operational, technical and functional maturity of the railway is progressively implemented;
5. System rollouts are achieved within the constraints imposed by the access requirements;
6. Roll-out costs are contained within acceptable levels.

The SUP will also interface with three network-wide PFI schemes, modernising the SSL infrastructure:

1. Communications - Connect PFI contracted with the consortium known as CityLink. It is responsible for supporting the radio system and providing reliable, high-capacity data links which can be flexibly deployed; a robust fibre-optic transmission network will also be installed. This will provide the SSL Railway with transmission links to support signalling control and communications systems;
2. Power - PFI contracted with EDF Energy Powerlink Limited (EEPL). The new trains and signalling to be delivered by the SSR programme will require more power than at present; therefore to accommodate the new trains, the high voltage distribution network

has to provide more power than is currently available. The PPP contract provides for LU to improve the existing power to meet these needs. With the integrated approach to the implementation of both programmes (SSL and Power PFI), this power upgrade will be in place prior to the full-scale introduction of new trains to the SSL Railway;

3. Ticketing - Prestige PFI contracted with the consortium known as Transys. There are few links between the Prestige PFI and the SUP, although there is a need for these to be identified and managed. An example is that many ticket gates in the central area use compressed air from the signalling air main for their motive power; this supply must be maintained despite the decommissioning of pneumatic signalling equipment (e.g. points) on the SSL Railway.

The SSL Programme is required to roll out an extensive upgrade of the 284km sub-surface network over the next 10 years. The upgrade will impact virtually all the assets on the current SSL network and will have a significant impact on the operation and maintenance of the railway during the intermediate and final state of the SSL Upgrade Programme. During the upgrade of the SSL railway, a number of interim railway system configurations are defined, as the new assets are delivered into service across the SSL railway network. In addition to the work needed to manage the procurement of the new rolling stock and signalling, additional enabling works are required in order to allow the new trains to operate on the line. The resulting work has been divided into four principal project portfolios as follows:

1. Train Systems portfolio – All work required for preparation, passenger service, availability, and maintenance of the first S Stock trains. This includes S Stock trains, necessary depot upgrades, DC traction upgrades including low loss conductor rail;
2. Track portfolio – All work required along the route to enable the S Stock trains to interoperate with existing rolling stock at line speed, includes train arrestors, gauging, maximum safe speed;
3. Immunisation portfolio – All work required in some areas of the SSL network to overcome safety concerns over the compatibility of the new trains with the existing train detection equipment. This includes signalling immunisation track circuit replacement, low voltage signalling power, cable route management and communications;
4. Station Enabling portfolio – All work required at stations along the route of the S Stock trains to interoperate with existing rolling stock in passenger service. This includes correct side door enable

(CSDE), selective door operation (SDO), OPO TT CCTV and platform extensions and modifications.

## 7.3 Challenge

Organisationally, both major upgrade programmes, SSL and VLU, are split into a number of layers; at programme layer, final integration into a railway system is undertaken whilst constituents are delivered by projects and the supply chain. The immediate integration into subsystems is realised by project portfolios of related projects.

For example, on the Victoria Line, the train is being delivered by Bombardier Passenger Group; the signalling control is being delivered by Westinghouse. Initial integration of the signalling onto train is carried out by Bombardier London Underground Projects.

Once the train-signalling subsystem integration is completed, the final integration of the subsystem is completed by the VL Upgrade Programme.

This is illustrated by Figure 7-2 below, where the blue colour indicates integration at project portfolio level and the red colour indicates the programme level integration.

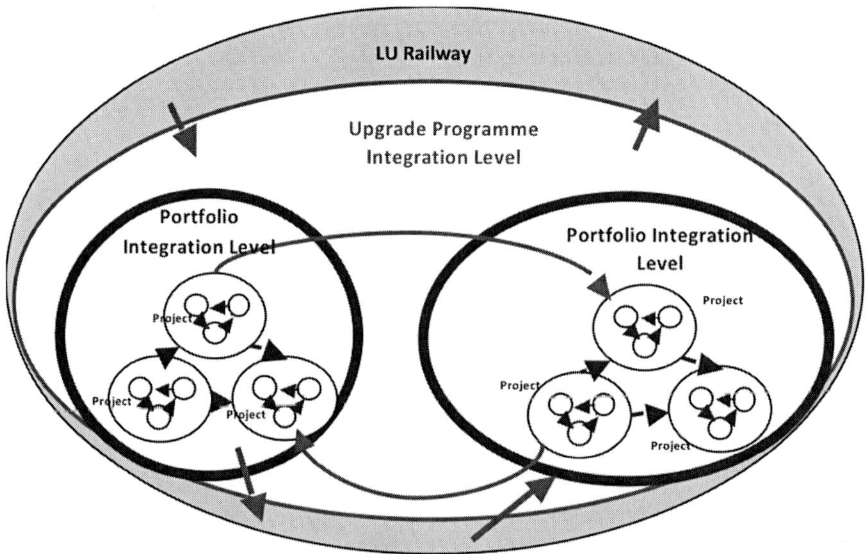


Figure 7-2: Organisational structure of the Upgrade Programmes and two integration layers

Integrating the safety arguments from a large number of parts and subsystems into an integrated railway system, and all delivered by different projects and a complex supply chain, into a series of holistic and coherent system safety cases (one for each of the main system architectures), is a challenge.

The complexity of the Upgrade Programmes necessitates an integrated approach to system safety management.

The author achieved this by ensuring that the various projects, suppliers and project portfolio's, and their specific engineering safety management processes, are guided by the common principles and practices (set out in the programme level Engineering safety Management Plan) and interact in such a way that the overall safety requirements and quantitative safety targets are met for each defined railway system configuration state.

As illustrated by Figure 7.3, there are three areas of focus when implementing the integrated safety analysis management on any complex project; alignment of processes, alignment and coordination of analysis and management activities, and harmonization and coordination of deliverables.

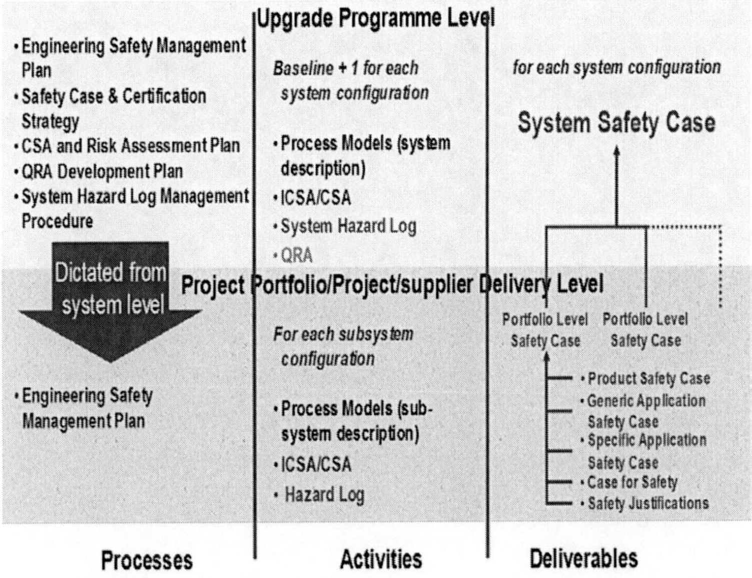


Figure 7-3: Integrated safety management process

When mapped onto the organisational structure of the upgrade programmes, this arrangement can be represented by Figure 7-4 below.

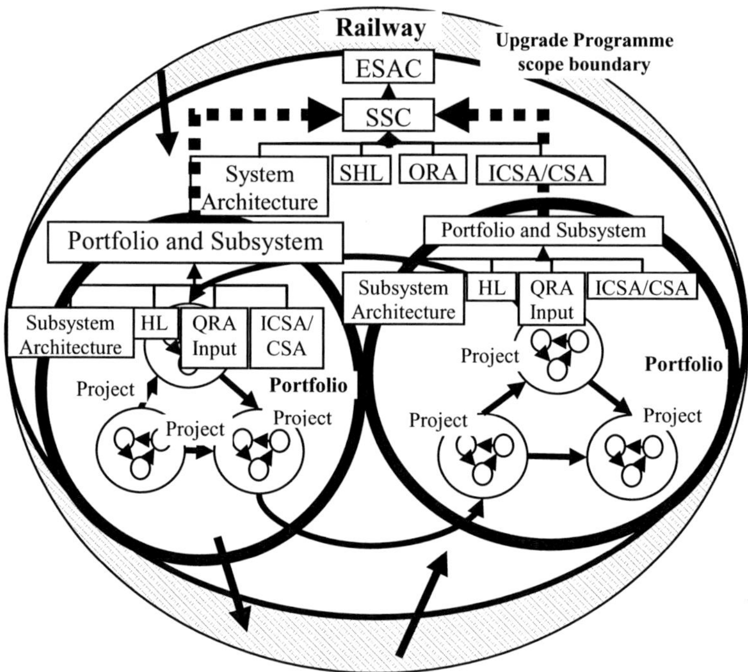


Figure 7-4: Integrated safety management process mapped onto organisational breakdown structure

Based around the safety fundamentals defined at the programme level, the following common engineering safety analysis and management fundamentals, are implemented across the project portfolios and key suppliers in order to enable an integrated approach to managing and demonstrating system safety for each of the configuration states:

1. Identification of a set of common core hazards. To ensure a common approach to hazard identification and analysis, it is essential that the same set of core hazards is used as a template structure for hazard log groupings across the programme and at system hazard log level across to the project portfolio-project level.
2. Use of a hierarchy of Process Models starting from the system level down to the delivered subsystem/constituent level in order to:

a. Define the scope of analysis at system and project/supplier level;
b. Facilitate common understanding of the scope of change to be introduced to the railway system;
c. Facilitate common understanding of interface boundaries between different projects/suppliers scopes of work;
d. Enable the traceability between different elements of the process, namely, QRA, Hazard Log and CSA in support of the completeness argument.

3. Use of the principles of Change Safety Analysis at programme and project level to identify hazards, mitigations and safety requirements. These have been explained in detail earlier.

4. Apportionment of Safety targets and quantification of requirements. To demonstrate that the risks resulting from the changes to the railway are tolerable, ALARP and no greater than the level of risk in the current Railway System, quantitative risk assessments are carried out, and a set of requirements related to the new hardware system constituents have been quantified.

   For the purposes of reporting against the risk performance, London Underground developed QRA models of the 20 most critical hazardous events, some time ago. The current London Underground Subsurface Lines and Victoria Line QRA (LU SSL and VL QRA) models are based upon an assessment of the risk of passenger fatalities from these hazardous events. These models have been reviewed with the aim of limiting the scope of the programmes QRA to only those interfaces and related hazards that are affected by the scope of the programmes' works as defined by reference to the Railway System Architecture.

   The subset of LU QRA models has been developed further to establish a baseline set of models representing the Current Safety Performance (CSP) of the SSL and VL prior to any changes introduced by the programmes' works. Predictions from these models have been used to set Tolerable Hazard Rates (high level safety targets) for both programmes.

   The set of CSP models has been further developed to reflect changes to the SSL and VL to be introduced by the programmes, for each stage system configuration and geographical area. As a result, a series of Interim Safety Performance (ISP) and Final Safety Performance (FSP) models of the SSL and VL have been developed. The models are used to quantify safety requirements for novel parts of the system and to confirm that the Tolerable Hazard

Rates (high level safety targets) have been met by each of the programmes for each system configuration during migration.

5. The use of a consistent hazard management process and hazard log structure across the programme, project portfolios and key suppliers including implementation of a Hazard Forum at programme and Interdisciplinary Hazard Review Group at the project portfolio level.

To demonstrate that the risks associated with the hazards introduced to the SSL and VL railways have been reduced to ALARP and tolerable, a robust hazard management process has been implemented. In addition, a clear relationship has been established between the Hazard Logs at the System (Programme), Subsystem (Project or Project Portfolio) and the Constituent (Supplier) level, and has in turn been documented thus aiming to adequately manage the hazards introduced to the operational railway as a result of the upgrade programmes.

The identified hazards are structured into two different classes for easy allocation and monitoring between system, subsystems and third party organisations where identified:

a. System boundary or interface hazards. These hazards reside at the interfaces between two subsystems (e.g. between rolling stock and signalling). Management of these types of hazards is the responsibility of the Systems Integration Team on each of the two programmes;

b. Subsystem level hazards. These types of hazards reside at the project and supplier level, and are the responsibility of the particular project portfolio, the project itself, or their key suppliers.

A Hazard Forum has been implemented at the programme level to oversee and address the transfer of hazards between the programme, projects and suppliers, and also to consider hazards that require effort from multiple projects to facilitate closure. The Hazard Forum is independent from the programme.

An Interdisciplinary Hazard Review Group [IHRG] was formed to address and close-out hazards at project portfolio, project and supplier level. The IHRG oversees and addresses the transfer of hazards between the programme, projects and suppliers, and also considers hazards that require effort from multiple projects to facilitate closure. The IHRG is independent from programme delivery.

The same hazard log management procedure and the same hazard log database structures are used across programme and project level. This ensures consistency between the system (programme) hazard log and project hazard logs.

6. Use of a common requirements management process, to support the traceability of safety requirements from programme to project, and finally, to supplier level and in the reverse direction. Safety Requirements (qualitative and quantitative) are built around the Core Hazards. This is because the only continuous "theme" relevant to the safety argument during the life of the programme of works is a set of the Core Hazards.

   Derivation of safety requirements was supported by the following sources:

   a. Railway Legislation and Regulations;
   b. LU Standards;
   c. European and International Standards;
   d. The PPP Service Contract;
   e. The Health and Safety Executive (HSE) Railway Safety Principles and Guidance;
   f. Safety Reports from past accidents in the UK railway industry;
   g. Preliminary Hazard Analysis – Safety requirements have been derived from the review of the mitigation measures identified during the PHA;
   h. Change Safety Analysis - Safety requirements have been derived from the review of the mitigation measures identified during the CSA.

   It has now been established that the above sources represent a set of core safety requirements that must be met by all project portfolios and suppliers. The safety requirements are generic, and include technical and process types. The requirements are applicable at all levels of the programme and for all systems and sub-systems put into operational use in the SSL and VL railway.

7. Linkage and consistency between engineering safety cases (safety arguments and evidences) produced at programme, project and supplier level in order to support the application for Consent to Operate, at key stages of the Upgrade Programmes. The underlying system safety argument is built around a set of safety goals derived from the Core Hazards identified in the System Hazard Log.

   The programmes make use of the process models to identify boundary interfaces, define and establish the linkages between the safety cases produced by programme, project and supplier. Identifying the responsibilities for each identified boundary interface enables the scope of each safety case to be fixed at programme, project and supplier level.

Adopting this approach ensures that there is full coverage of all safety issues arising from the SUP programme works, that there are no gaps in the analysis, but as well, that there is no duplication of effort.

The technical safety report section of the safety cases are based around the Core Hazards, at project and programme level. Central to demonstrating safety is the use of GSN to outline the safety argument and set out the evidence supporting the case for safety.

In constructing a GSN based safety argument for an overall system safety case, the safety cases of individual sub-systems within the system are used to supply the evidence. The core hazards are a constant for all parts of the system and for all configurations of the project, and their use forms the goals of the safety justification for each railway system configuration as the project is progressing and facilitates the development of consistent and reusable safety arguments. This includes the use of Goal Structuring Notation (GSN), to:

a. Facilitate the identification and management of a hierarchy of safety arguments and evidence throughout the programme;
b. Facilitate the development of railway system safety arguments based on the safety arguments for individual subsystems (developed by projects or suppliers);
c. Enable the reuse of safety arguments from one stage to another.

## 7.4 Identification of a set of common core hazards

The Top Events of the LU QRA were used as the starting point for identification of the Core Hazards. Only those QRA Top Events were included in the version that was believed to be affected by the changes to the railway were included within the scope of the upgrade programmes.

As already stated, the LU QRA is of very narrow scope, and in order to confirm the completeness of the core hazards set and its scope, it was necessary to identify and fill any gaps (for example, workforce related risk and causes which do not result in fatality).

A review was carried out using the following complementary sources of railway risks to complete the list of LU QRA top events:

1. Railway Safety and Standards Board (RSSB) Safety Risk Model;
2. Victoria Line Upgrade (VLU) System Safety Case Core Hazards;
3. Existing SSR System Hazard Log hazard groupings;
4. West Coast Route Modernisation Current Safety Performance of the Railway QRA model.

Each of the QRA Top Events was reviewed for possible omissions highlighted by the other sources. Where appropriate, the scope of the Core Hazard mapped to the QRA Top Event, was broadened.

Where there was no suitable existing LU QRA Top Event, a new Core Hazard was defined. The Core Hazards were named to maintain an obvious link to any relevant QRA Top Events. The names of Top Events defined in terms of accidents were modified to better reflect the definition of a Hazard e.g. "the potential to cause harm".

| Impacts on | Core Hazard | Definitions |
|---|---|---|
| Passenger, Workforce, Member of Public (MOP) | Potential for collision between trains | This Core Hazard groups causes, arising from or impacted by the scope of works, which effect the safe separation between trains.<br><br>This Core Hazard is primarily expected to affect Passengers and Workforce. There is potentially a secondary effect on MOP, were a collision could result in "something hazardous" going beyond the railway boundary. |
| Passenger, Workforce, Member of Public | Potential for collision with object | This Core Hazard groups causes, arising from or impacted by the scope of works, which affect the likelihood of objects or animals "find their way" on or near the running railway such that they could make contact with a passing train.<br><br>This Core Hazard includes those occurrences where the train:<br>• Is incompatible with the structure gauge;<br>• May collide with buffers.<br><br>This Core Hazard excludes:<br>• Instances of objects on the track |

| Impacts on | Core Hazard | Definitions |
|---|---|---|
| | | causing fires;<br>• Collision with objects which cause derailment.<br>This Core Hazard is primarily expected to affect Passengers and Workforce. There is potentially a secondary effect on MOP, were a collision to result in "something hazardous" going beyond the railway boundary. |
| Passenger, Workforce, Member of Public | Potential for derailment | This Core Hazard groups causes, arising from or impacted by the scope of works, where the relationship between the track and the train is compromised such that the train may be derailed. Examples of possible causes include: over-speeding of the train, track degradation outside safe limits, faults at switches and crossings, signalling failures and objects on the line.<br><br>This Core Hazard is primarily expected to affect Passengers and Workforce. There is potentially a secondary effect on MOP, were a derailment to result in "something hazardous" going beyond the railway boundary. |
| Passenger, Workforce | People-train incident at the platform interface | This Core Hazard groups causes, arising from or impacted by the scope of works, which affect people at the platform train interface including, but not restricted too:<br>• Entering or alighting from trains;<br>• Falling off platforms;<br>• Being struck or run over by train (station areas only);<br>• Crossing the lines at station (Where authorised only);<br>• Opening and closing of carriage doors.<br><br>This Core Hazard excludes Passengers and Workers who deliberately access restricted track areas. These events are within the scope of Unauthorised access to track. |

| Impacts on | Core Hazard | Definitions |
|---|---|---|
| | | This Core Hazard has been extended to also include slip, trip and fall Causes arising from or impacted by the changes in platform areas undertaken within the scope of works.<br>This Core Hazard is expected to affect Passengers and Workforce only. |
| Passenger, Workforce | People-train incident on the train | This Core Hazard groups causes, arising from or impacted by the scope of works, which affect people on-train due to train movement: This includes, but is not restricted too:<br>• People protruding beyond train gauge during movement;<br>• Loss of train compartment integrity (e.g. Carriage separation, broken windows);<br>• Falls due to train lurching, jerking or rapid deceleration.<br><br>This Core Hazard excludes a collision with another train, derailment, collision with an object. (All of these are covered by other Core Hazards).<br><br>This Core Hazard is expected to affect Passengers and Workforce only. |
| Passenger, Workforce, Member of Public | Potential Electrocution | This Core Hazard groups causes, arising from or impacted by the scope of works, which affect the safe separation of people from live electrical power supplies.<br><br>This Core Hazard is closely linked to Arcing but is kept separate because the nature of any resulting accident will be different. However the two Core Hazards may be combined together for analysis where it is pragmatic to do so.<br><br>This Core Hazard may affect Passengers, Workforce and Members of the Public. |

| Impacts on | Core Hazard | Definitions |
|---|---|---|
| Passenger, Workforce, Member of Public | Arcing | This Core Hazard groups causes, arising from or impacted by the scope of works, which affect arching of electrical power supplies (particular traction power).<br><br>This Core Hazard is closely linked to Potential Electrocution but is kept separate because the nature of any resulting accident will be different. However the two Core Hazards may be combined together for analysis where it is pragmatic to do so.<br><br>This Core Hazard may affect Passengers, Workforce and Members of the Public. |
| Passenger, Workforce | Potential for Train Fire | This Core Hazard groups causes, arising from or impacted by the scope of works, which affect the potential for on-train fires.<br><br>This Core Hazard is expected to impact on-train Passengers and Workforce. |
| Passenger, Workforce, Member of Public | Potential for Station Fire | This Core Hazard groups causes, arising from or impacted by the scope of works, which affect the potential for station fires.<br><br>This Core Hazard may affect Passengers, Workforce and Members of the Public. |
| Passenger, Workforce | Potential for Trackside Fire | This Core Hazard groups causes, arising from or impacted by the scope of works, which affect the potential for trackside fires. For the purposes of this Core Hazard trackside shall include both open line-side and sub-surface.<br><br>This Core Hazard may affect Passengers, Workforce and Members of the Public. |
| Passenger, Workforce | Inadequate ventilation (On train) | This Core Hazard groups causes, arising from or impacted by the scope of works, which affect the on-train air quality.<br>This Core Hazard excludes fires.<br><br>This Core Hazard may affect Passengers and Workforce. |

| Impacts on | Core Hazard | Definitions |
|---|---|---|
| Passenger, Workforce, Member of Public | Unauthorised access to track | This Core Hazard groups causes, arising from or impacted by the scope of works, related to unauthorised access to track.<br><br>This Core Hazard may affect Passengers, Workforce and Members of the Public. |
| Workforce | Failure to protect workforce on track from train movements | This Core Hazard groups causes, arising from or impacted by the scope of works, related to workforce when on or about the track. This includes but is not limited to:<br>• When acting as lookout or hand signaller;<br>• When working on or about the track;<br>• When authorised to walk on the track;<br>• Flying objects or out-of-gauge parts of a train.<br><br>This Core Hazard will impact on the Workforce only. |
| Workforce | Workforce Occupational Health and Safety | This Core Hazard is intended to encompass all causes, arising from or impacted by the scope of works, related to Workforce occupational health and safety which fall outside the scope of all other Core Hazards.<br><br>This Core Hazard will impact on the Workforce only. |
| Passenger, Workforce | Potential for flood | This Core Hazard is intended to encompass all Causes, arising from or impacted by the scope of works, relating to the potential for flooding.<br><br>This Core Hazard may affect Passengers, Workforce and Members of the Public. |
| Passenger, Workforce, Member of Public | Potential for Structural failure | This Core Hazard is intended to encompass all Causes, arising from or impacted by the scope of works, relating to the failure of structures.<br><br>This Core Hazard may affect Passengers, Workforce and Members of the Public. |

| Impacts on | Core Hazard | Definitions |
|---|---|---|
| Passenger, Workforce | Evacuation from trains | This Core Hazard is intended to encompass all Causes, arising from or impacted by the scope of works, relating to the evacuation of trains.<br><br>This Core Hazard may affect Passengers and Workforce. |

**Table 7-1: SSL Core Hazards definition**

## 7.5 Information gathering, conceptualisation and representation (System Analysis)

To assist the understanding of the interaction between the various systems and elements which comprise the overall system, for each Core Hazard (or a group of Core Hazards), process models have been developed based on the Railway System Physical Architecture, and for a relevant Railway Configuration, with the aim to:

1. Identify the system boundaries;
2. Identify boundary interfaces and their associated change through migration stages;
3. Identify owners of changes to each boundary interface (i.e. supplier, project or programme);
4. Identify safety functions of the system;
5. Identify the subset of the system relevant to each safety function;
6. Focus attention on the functional interactions between the elements relevant to each safety function;
7. Focus Change Safety Analysis on the elements of the railway that are subject to change;
8. Establish a common reference environment for Safety Analysis and Derivation of Safety Arguments across the programmes;
9. Ensure that there is no duplication of effort in the safety analysis carried out at programme, project and supply chain.

Process models have been developed with particular focus on the identification and establishment of boundary interfaces and their associated physical, functional or operational change. Responsibilities for each change, to any of the interfaces, have been identified and assigned. The scope of the safety analysis is clearly determined by the use of process models, ensuring that there is no duplication of effort in the safety analysis carried out at programme, project and supply chain.

More importantly, the use of process models provides focus for the analysis on the elements of the railway, which are subject to change by the upgrade programme scope of work. A common approach to the conceptualisation of the system was adopted in support of the two upgrade programmes, VLU and SSL.

For the purpose of this book, an example of the process model depicting the same family of "Train accidents" from both programmes has been used. The "Train accidents" comprise collision between trains, collision of trains with buffer stops or movement arresters, derailments, and the collision of trains with objects.

The model shown here represents the situation of the first new trains running interspersed with existing trains, and the line still being controlled from the existing control room and existing interlocking. For simplicity, the existing trains and their relation to the track and existing system have not been shown. The safe behaviour of these elements of the system is taken as already proven.

The existing signalling system acts on the track in the sense that it drives the points. It receives information from the track in the form of track circuit inputs and points detection. It also provides energy to lineside signals and receives information in the form of lamp proving.

Specifically for VLU, the new signalling system receives information from the existing signalling system, and provides information to control the train via its traction control and braking systems. It should be noted that the element labelled "Signalling System (new)" includes the on-train parts of the signalling system as well as the lineside parts.

The controller controls and monitors the movement of trains via the Service Control Centre system and its links with the signal system, and the line controller and driver can speak via a voice communications link. In support of the Victoria Line safety analysis, the process models have been developed at the programme/railway system level only. An example from VLU on Figure 7-5 is the process model covering the following core hazards; "Potential for collision between trains", "Potential for collision with object" and "Potential for derailment" in one process model. An example of interface descriptions for this process model is in Table 7-2 below.

This methodology was chosen because by the time the CSA and ESAP have been implemented, the supply chain and project activities have advanced significantly further than the railway system level activities and, therefore, it was not practicable to develop the hierarchy of process models. Instead, allocation of responsibilities for development of safety arguments was implemented through the mapping of the VLUP System Architecture interfaces (SCID and O&Idb interfaces) to the supplier's interfaces matrix.

| Process Model Ref | IF Ref | Interface (from SCID) Description |
|---|---|---|
| TA07 | IF0409 | 09TS Train provides discrete status indications to the Train Operator via the Drivers HMI in the cab. |
| TA25 | IF1352 | Track Lubricator provides Lubrication using Mechanical Device (method) via Mechanical (medium) to Trackwork |

**Table 7-2: VLU Train Movement Accident interfaces description**

On SSL, the process models were developed at the programme level initially, and subsequently at the project level, resulting in a hierarchy of the process models providing two levels of detail commensurate to the level of detail required to support the engineering safety analysis and management activities.

As depicted by Figure 7-6, the System Level Process model has been further decomposed for each project portfolio scope of works in order to provide the relevant detail in support of safety analysis at each of the two levels.

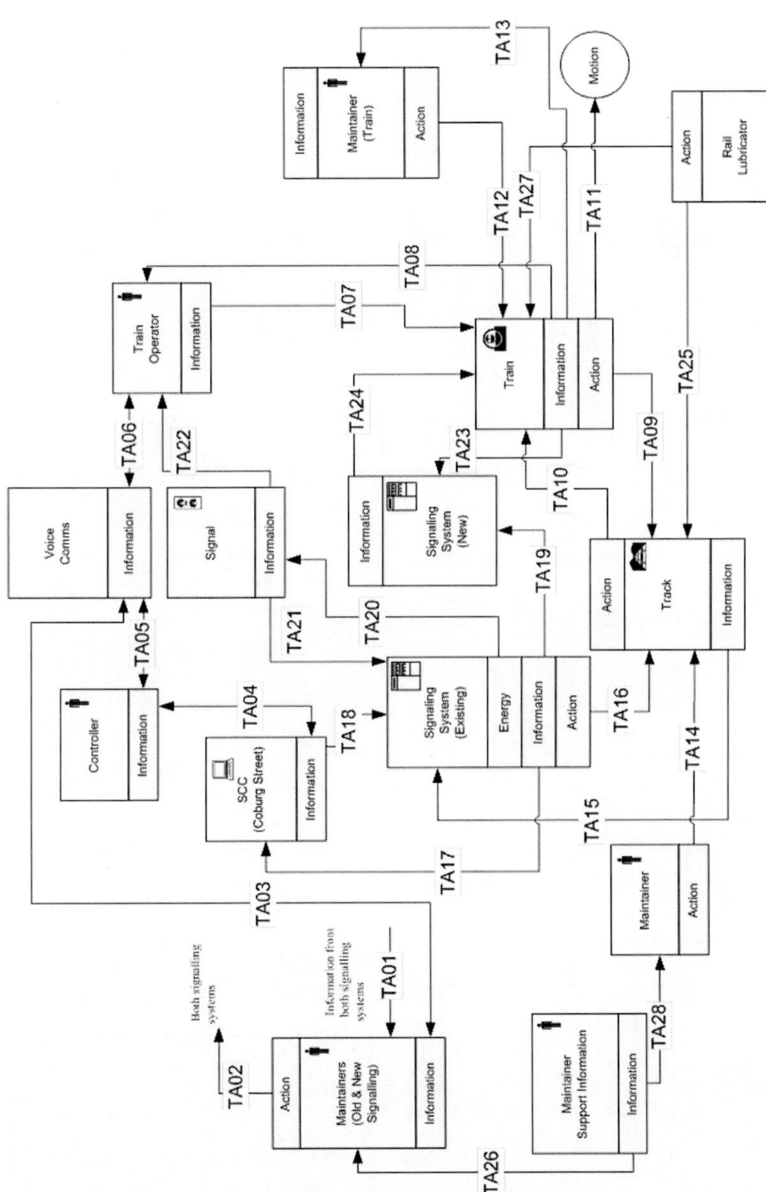


Figure 7-5: VLUP Train Movement Accidents process model

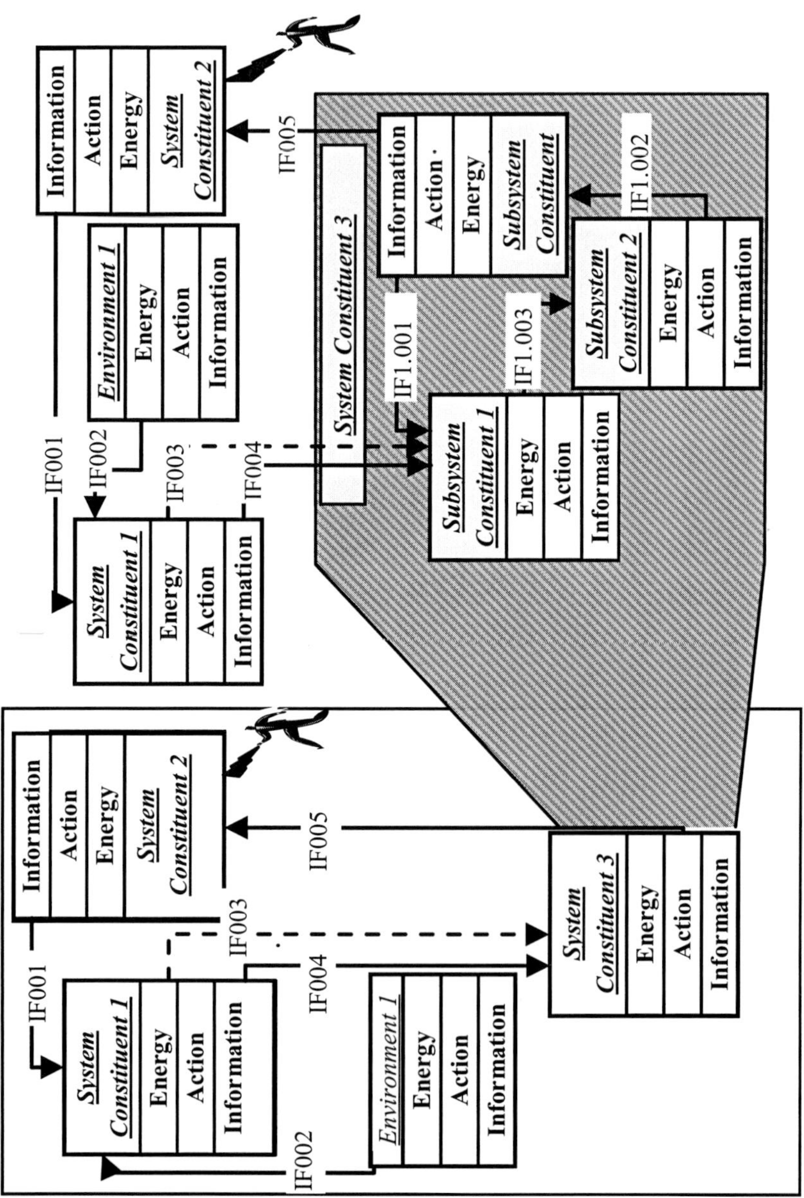


Figure 7-6: Decomposition of process models

An example below, Figure 7-7, represents the system level SSL "Train Movement Accidents" process model. This process model has been decomposed further into a number of process models for the "Immunisation Portfolio"; "Rolling Stock Portfolio", etc. In this example, system components indicated by coloured boxes have been decomposed in order to reflect the scope of the "Immunisation Portfolio" works (see Figure 7-8). An example of the process models' interfaces descriptions and hierarchical decomposition is presented in Table 7-3 below.

<table>
<tr><td colspan="5">SSL System level Process Model (Figure 7-7)</td></tr>
<tr><td>Process Model Interface Id</td><td>Upstream Object</td><td>Downstream Object</td><td>Interface Description</td><td>Change</td></tr>
<tr><td>TM07</td><td>Current Signalling System</td><td>Train Stop</td><td>I/L command to actuate train stop movement</td><td>Train Stop moved.</td></tr>
<tr><td colspan="5">Project Portfolio level Process Model (Immunisation Portfolio) (Figure 7-8) with mapping to decomposed interfaces</td></tr>
<tr><td>TMA29/ TMA65</td><td rowspan="2">Current Signalling System object controller</td><td>Train Stop via Auxiliary Signal</td><td rowspan="2">I/L command to actuate train stop movement</td><td rowspan="2">Train Stop moved.</td></tr>
<tr><td>TMA30/ TMA64</td><td>Train Stop via Main Signal</td></tr>
</table>

**Table 7-3: SSL Train Movement Accident interfaces description**

The "Immunisation Portfolio" Level of the "Train Movement Accidents" process model provides more detail that is relevant to the scope of the change to be introduced by this portfolio. On Figure 7-8, the decomposition of relevant system components is indicated by highlighted boxes.

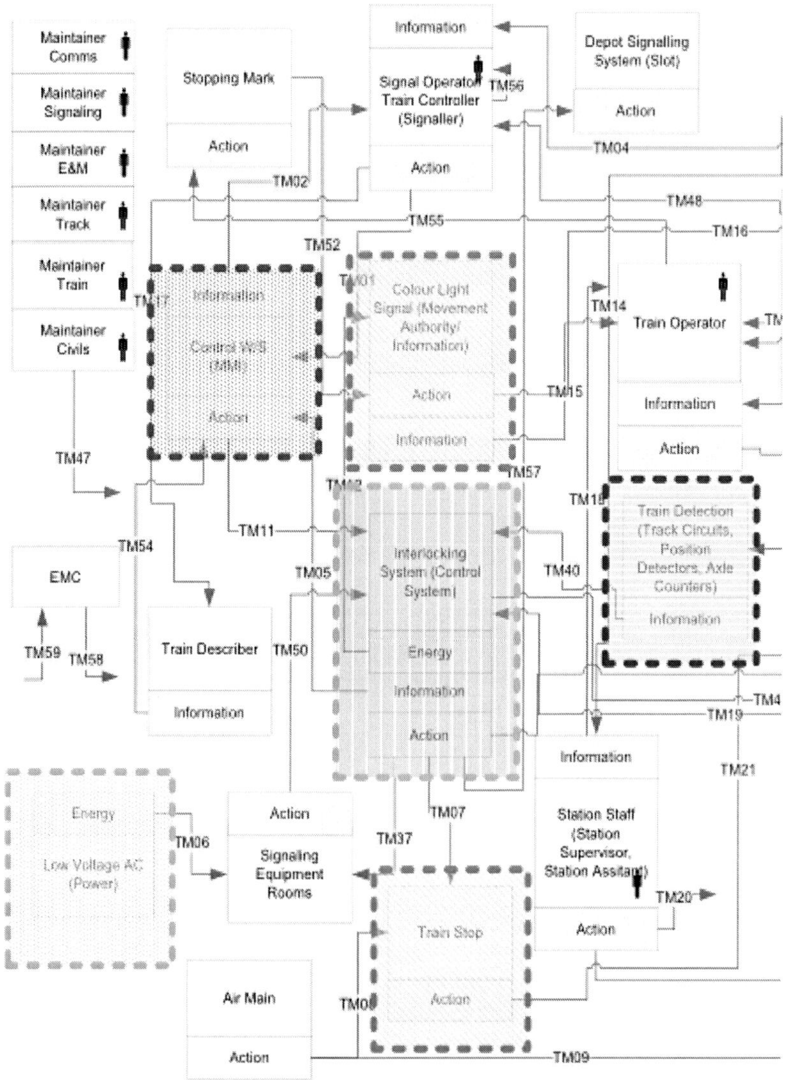


Train accidents comprise collision between trains, collision of trains with buffer stops or movement arrestors, derailments, and collision of trains with objects

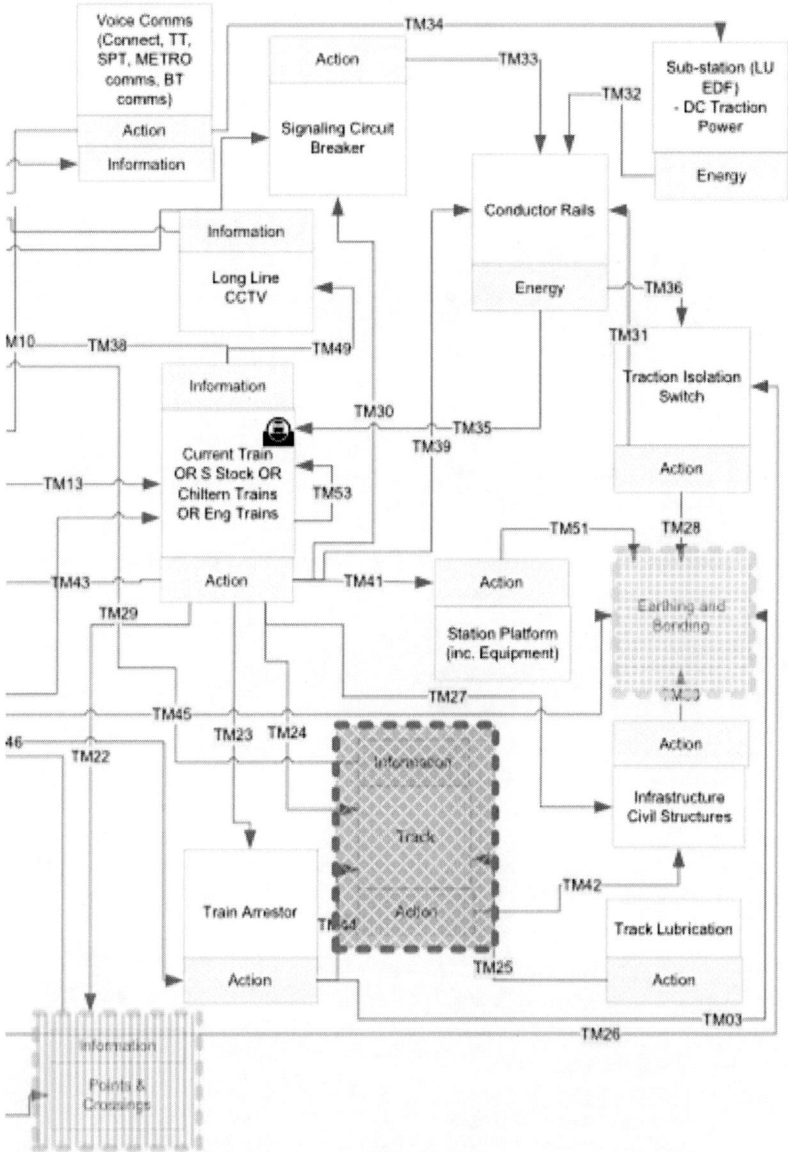


Figure 7-7: SSL System Level “Train Movement Accidents” process model

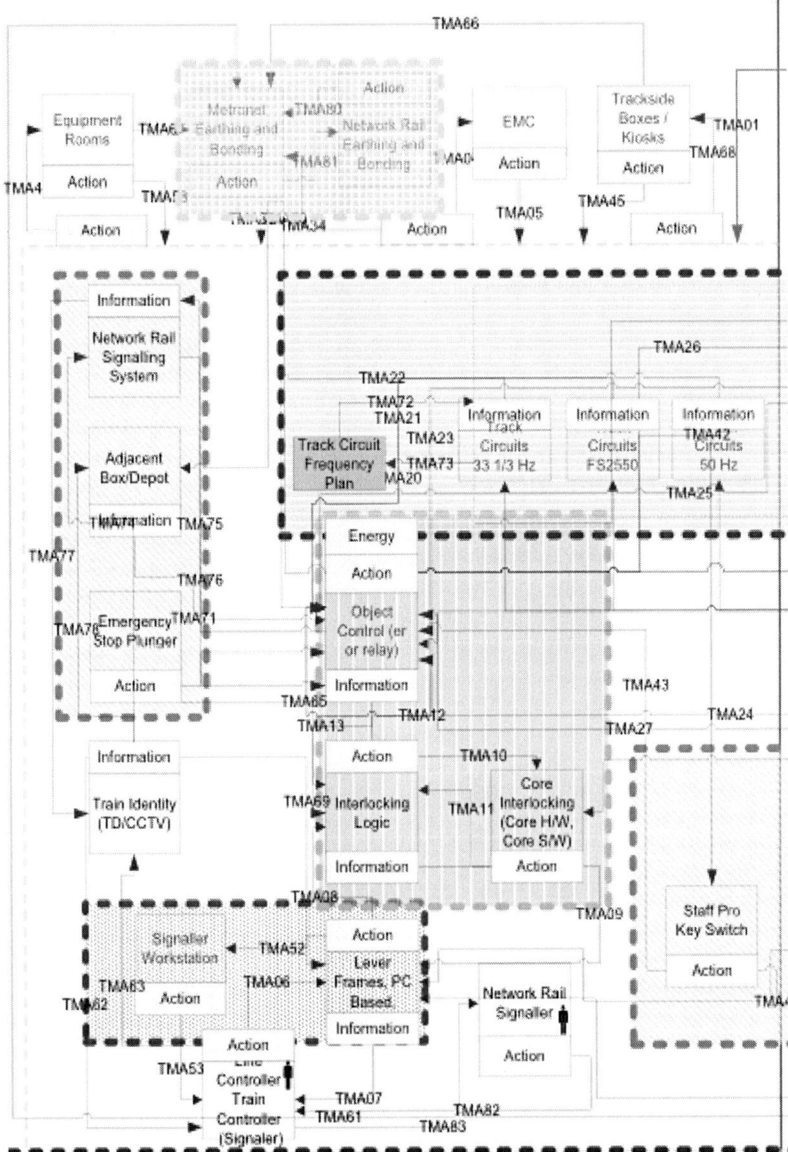
TMA66
Equipment Rooms
Action
Metronet Earthing and Bonding
Action
Network Rail Earthing and Bonding
Action
TMA80
TMA81
EMC
Action
Trackside Boxes / Kiosks
Action
TMA01
TMA68
TMA4
TMA45
TMA05
Action
Action
Action
Action
TMA34
Information
Network Rail Signalling System
Adjacent Box/Depot
Information
TMA75
TMA77
TMA76
Emergency Stop Plunger
TMA71
Action
TMA26
TMA22
TMA72
TMA21
TMA23
TMA73
TMA20
Track Circuit Frequency Plan
Information
Track Circuits 33 1/3 Hz
Information
Circuits FS2550
Information
TMA42
Circuits 50 Hz
TMA25
Energy
Action
Object Control (er or relay)
Information
TMA65
TMA12
TMA13
TMA43
TMA24
TMA27
Information
Train Identity (TD/CCTV)
Action
TMA10
Interlocking Logic
Information
TMA69
TMA11
Core Interlocking (Core H/W, Core S/W)
Action
TMA08
TMA09
Staff Pro Key Switch
Action
Signaller Workstation
Action
TMA52
TMA06
TMA63
Action
Lever Frames, PC Based
Information
Network Rail Signaller
Action
Action
TMA53
Line Controller
Train Controller (Signaler)
TMA07
TMA61
TMA82
TMA83

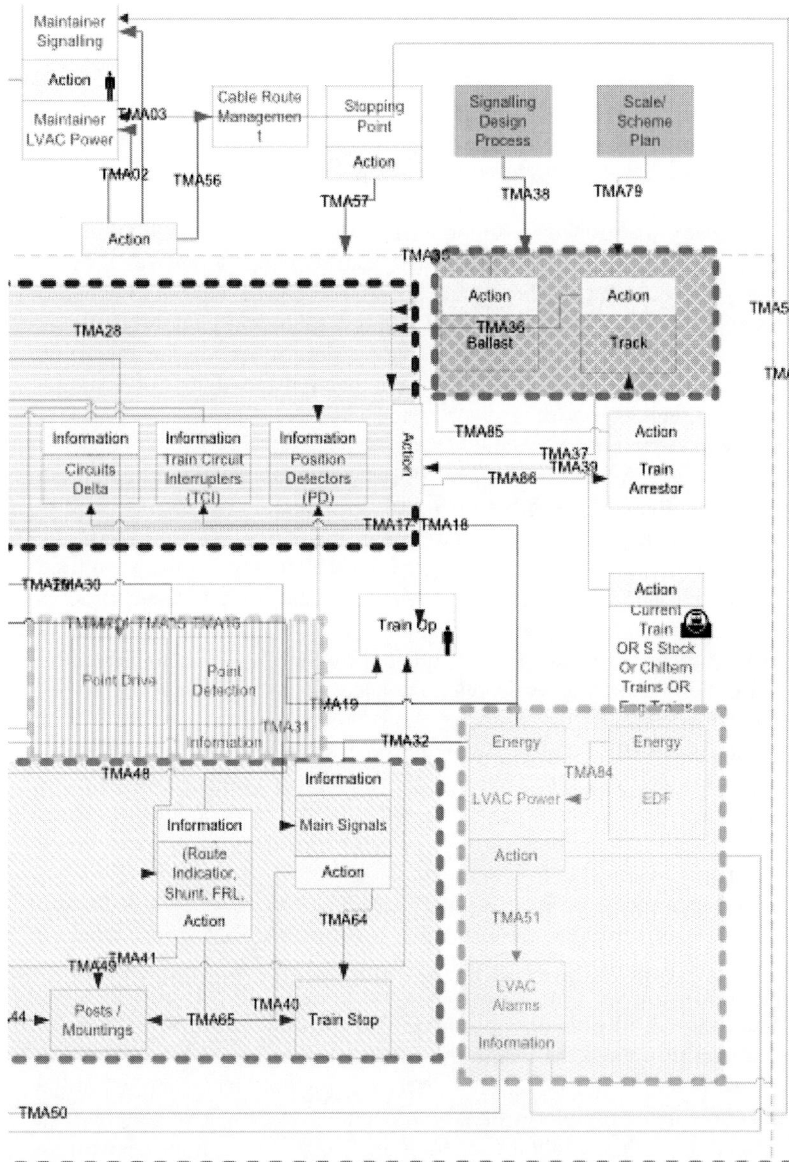


Figure 7-8: SSL "Immunisation Portfolio" Level "Train Movement Accidents" process model

## 7.6 Information processing

### 7.6.1 Change Safety Analysis: hazard identification, analysis of causalities, consequences, options and impact analysis and risk assessment

As already mentioned in support of further hazard identification, for each Core Hazard a Process Model has been developed and was used to structure the Change Safety Analysis as described earlier. For each interface, the ICSA class is identified and recorded, and a reference to the relevant core hazard is kept as well as the description of the identified cause. All of the identified causes have been mapped to QRA model elements as well. This activity supported later quantification of some of the safety requirements as detailed later in this book.

The comparable risk assessment process was also successfully applied (pre and post implementation of the mitigation) and the results of it used in the safety case arguments as detailed later.

In order to assess the correctness and sufficiency of the identified mitigation measures, the ALARP review process followed the process depicted in scheme provided in Table 7 4. This ranking was used to confirm whether the residual risk presented by the hazard, as mitigated by the mitigation measures, is acceptable in accordance with the acceptability matrix presented in Table 7 5 .

It should be noted that the depicted process relies upon the collective expert knowledge and judgement of the assembled workshop participants to determine whether all relevant mitigations have been recorded and make a first-pass judgement on whether any additional mitigations could be reasonably practicable to implement.

To aid the process of assessing the risk of the hazard as a whole as well as adding transparency to the classification process, the author added an additional step to the process depicted in Figure 7-9; the process was applied (by using the same classifications, as in Table 7-4) at the Causal level. It was not always possible to assign a classification at the Causal level, and it was recorded where this was the case. The risk acceptability criteria shown in Table 7-5 are not directly applicable when applied at the Causal level. For example, it may be acceptable to have a worsening of a minor cause as this might be counter balanced by an improvement in another cause to give no effect on the hazard as a whole. This is particularly relevant at the overlay stages (when new and old trains are to be running at the same time, old trains relying on the old signalling system and new trains on the new DTG-R signalling) as there may be an increase in risk for some causes associated with operational arrangements for trialing (e.g. driver distraction due to additional personnel in cab). Qualitative ranking has been undertaken on a per-hazard basis in accordance with the classification scheme provided in Table 7-4. This ranking was used to confirm whether the residual risk presented by the hazard, as mitigated by the mitigation measures, is acceptable in accordance with the acceptability matrix presented in Table 7-5.

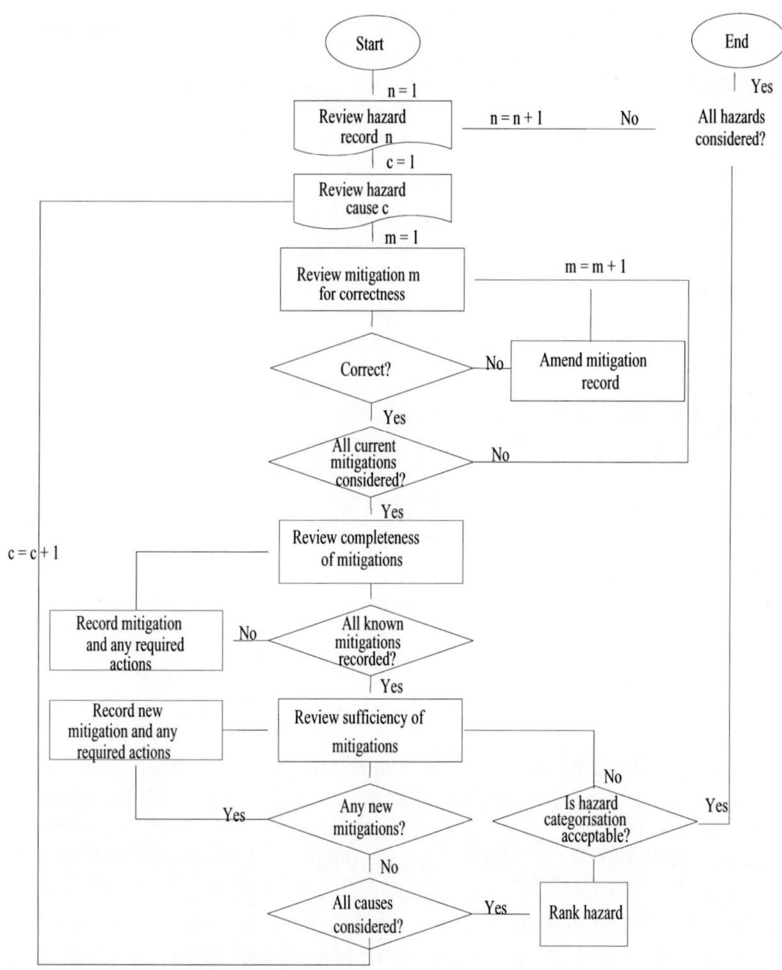


Figure 7-9: ALARP Review Process

| | Qualitative Ranking | Definition |
|---|---|---|
| Classification 1 | BETTER | The risk is assessed as significantly better (lower) than would be achieved with standards or other established authoritative good practice being used |
| | WORSE | The risk is assessed as significantly worse (higher) than would be achieved with standards or other established authoritative good practice |
| | COMPARABLE | The risk is assessed as neither Better nor Worse than would be achieved with standards or other established authoritative good practice |
| Classification 2 | MINOR | There are other hazards associated with the VLU Programme that have a risk at least an order of magnitude greater for the group which is exposed to them in the relevant part of the VLU |
| | MAJOR | There is no other hazard associated with the VLU Programme that has a risk at least an order of magnitude greater |

**Table 7-4: Qualitative Ranking**

| | BETTER | COMPARABLE | WORSE |
|---|---|---|---|
| MINOR | Acceptable | Acceptable | Not Acceptable |
| MAJOR | Acceptable | Further Review | Not Acceptable |

**Table 7-5: Risk Acceptability**

The author used the results of the review, i.e. the options and impact analysis, in support of system safety case arguments.

On VLU, process models at the system level were subjected to ICSA and CSA. The analysis was conducted to identify the system level causes, defence/protection measures, and consequences related to changing interfaces identified to be between the scopes of work of different suppliers and/or projects. An example of the outcome of the process is presented in Table 7-6 below.

On SSL, process models at the system level were subjected to ICSA and CSA first. The system level analysis was conducted in order to identify system level causes, defence/protection measures and consequences related to the changing interfaces identified to be between the scopes of work of project portfolios or where the change is introduced by mode, rather than one project portfolio.

Following on from that, a project portfolio/ subsystem level CSA was conducted, aiming to identify the sub-system or constituent level causes, defence/protection measures and consequences related to interfaces being changed at this level. An example of the ICSA/CSA record related to the SSL System Level Analysis is presented in Table 7-7 below. As illustration of the hierarchy of the Change Safety Analysis a change to interface TM07 is taken as an example.

This change, displacement of Train Stops, is being delivered by the Immunisation Portfolio Project. Part of a railway signalling system, a train stop or trip stop (sometimes called a tripper) is a train protection device that automatically stops a train if it attempts to pass a signal when the signal aspect and operating rules prohibit such movement, or if it attempts to pass at an excessive speed. As part of the changes implemented by the portfolio, the power supply to the signalling system is being upgraded and the train stops are being relocated. Results of the related CSA are presented in the Table 7-8 below.

| Process Model Ref | ICSA (Change) Class | Core Hazard | Cause | | Mitigation | | Risk Assessment Classification | | | | Mapping to QRA model | |
|---|---|---|---|---|---|---|---|---|---|---|---|---|
| | | | | | | | pre | | post | | | |
| | | | Id | Cause Description | Id | Mitigation Description | 1 | 2 | 1 | 2 | Id | Description |
| TA25 | 2 | Derailment | 33459 | Insufficient rail lubrication (both T1 and T2 present) | 1564 | Effective WRI management regime implemented covering lubrication, track inspection and maintenance. | Comparable | Minor | Comparable | Minor | LUBCATN | Poor or inadequate lubrication |

| | | | | | | | | | | | |
|---|---|---|---|---|---|---|---|---|---|---|---|
| TA07 | 1 | Collision between trains | 1393 | Train Operator error -09TS train operator will be presented with a higher target speed in PM mode in comparison to 67TS. The train operator may over speed when returning to a 67TS and the signalling system will not trip the train at 22mph in Coded Manual. | 6543 | 09 train stock dedicated drivers (if necessary, additional driver briefing). | Comparable | Minor | Comparable | Minor | TO_OVRS_CM | TOp ignores rule and exceeds 22 mph in PM |

**Table 7-6: VLU Train Movement Accident-CSA Output Result**

| Process Mod Ref | ICSA Class | Core Hazard | Cause | | Mitigation | | Risk Assessment | | | |
|---|---|---|---|---|---|---|---|---|---|---|
| | | | | | | | pre | | post | |
| | | | Id | Cause Description | Id | Mitigation Description | 1 | 2 | 1 | 2 |
| TM07 | 1 | Potential Collision Between Trains | 247 | Loss of Tripcock protection. Trainstop failed in up position while clear aspect is given. (Trainstop fails in up position causing damage to the tripcock when it passes the trainstop at speed, such that the tripcock is then not able to protect the train at a subsequent trainstop) | 151 | • Signalling protection (Alternate Red Light Circuit driving red lamp)<br>• Signalling Design Process Implemented by competent people;<br>• Testing by Competent people;<br>• Appropriate concessions generated & approved. | Comparable | Minor | Comparable | Minor |

**Table 7-7: SSL Train Movement Accident-System Level CSA Output Results**

| Process Model Ref | CSA (Change) Clas | Core Hazard | Cause | | Mitigation | | Risk Assessment | | | |
|---|---|---|---|---|---|---|---|---|---|---|
| | | | | | | | pre | | post | |
| | | | ID | Cause Description | ID | Mitigation Description | 1 | 2 | 1 | 2 |
| TMA29/TMA65 & TMA30/TMA64 | 1 | Collision between trains | 008 to 010 | Signalling System Unavailable – Loss of supply/incorrect or inadequate supply to signalling equipment: Overall safety risk is considered comparable, some sites are being considered for improvements in availability. Historical discussion around the design methodology. Some changes are recommended however, the project believes that a business case must be generated to enable changes via the ECB. | 1620 | Recommendation that FRACAS process should be maintained until confidence in failure rates is achieved. | Comparable | Minor | Comparable | Minor |
| TMA29/TMA65 and TMA30/TMA64 | 1 | Collision between trains | 013-026 and 034 | RSF, WSF Failures: The overall system EMC has been achieved. Maintenance is being undertaken by the project. Signalling and LVAC training and maintenance is not tested as comparable to having the maintenance organisation undertaking management of maintenance activities. Maintainer will follow the agreed maintenance process. FS2550 Failures (FRACAS reported); Effect is only availability and associated secondary risks (Spads etc). Further discussion on the ballast monitoring regime – no new issues raised. | 1564 | Training completion and Manual provision must be completed for maintenance staff. Wherever ballast needs to be changed, this has been completed or timescale for change undertaken. Caveat: monitoring is either deemed sufficient for M0 or completed. | Worse | Minor | Comparable | Minor |

**Table 7-8: SSL Train Movement Accident-Project Portfolio Level CSA Output Results**

### 7.6.2 Restrictions management

It is important to distinguish between the Restrictions, Dependencies and Assumptions. A dependency is an agreement between involved parties that something will be in place before the operation of the change. An assumption is a statement about the system or the rest of the world, including people and organisations with which it will interact, as well as the physical system and the environment. If the above definitions are accepted, both assumptions and dependencies need to be managed and controlled by the project or undertaking, but do not necessitate safety analysis. The restrictions however need to be treated in same way as defence/protection measures.

A Restriction is a condition or limitation imposed on the system that must be respected, after the change is put into operation, for the system to remain safe and operable. These can be categorised as:

1. Permanent restrictions – imposed on the system by limits of the operational envelope of the system within its environment;
2. Temporary restrictions – imposed on the system due to interim shortcomings of the system design or implementation.

With the increased complexity of the system, in the early stages of a new or a changed system introduction, the number of restrictions is likely to rise; in particular, the number of temporary restrictions is likely to be relatively high. On VLUP, as part of the safety analysis in support of the first stage of the project implementation (trial running of one new train in normal operational hours, controlled by the new signalling system), the author invented the following process (Figure 7-10) to identify and analyse all the restrictions. The process is structured as follows:

A. Review and Categorisation: The aim of this is to clearly scope out the boundary of review. This activity is carried out in two steps:

Step 1: Characterisation of inputs into Assumptions, Dependences and Restrictions/Caveats.

Step 2: Identification of causes of identified restrictions and confirmation of completeness and correct mapping to related hazards.

B. A preliminary assessment of the impact of causes of restrictions/caveats and identification of Mitigation options:

The aim of this part of the review is to prioritise and carry out the identification of options to mitigate identified safety issues and concerns.

This is carried out in two steps:

Step 1: Review the restrictions applying the criteria P1: Is a cause/issue leading to the restriction:

1. A direct cause to an accident?
2. A secondary cause to an accident?
3. Not safety related?

If a cause/issue does not lead to a safety accident then it is recorded and no further action taken.

If a cause/issue is potentially a direct cause of a safety accident, it is flagged and subject to further review focussing on these first. If a cause/issue is potentially a secondary cause of a safety accident it is subjected to further review after completing the review of direct causes.

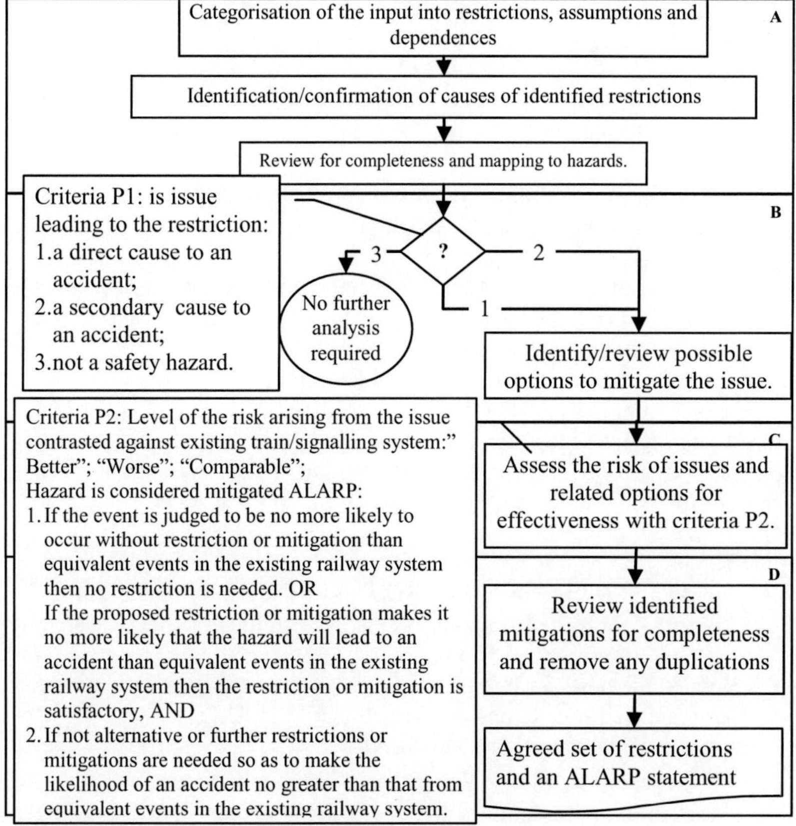


Figure 7-10: Restrictions identification and analysis process

*[Example: Causes of Restrictions on the VLU at the stage of the project are typically a fault in the implementation of the software product, a fault in the geographic data or a fault in the product design. For example, Manned Auto Operation, Unimpeded, into platform two is caused by geographic data errors in the map or APRs incorrectly located / programmed at the ramp to the depot. The following hazards have been identified as relevant:*

1. *Inappropriate separation between trains leading to a collision;*
2. *Possible derailment due to wrong movement of train over unsecured or out of correspondence points, or movement of points under the train;*
3. *Possible derailment due to over speed;*
4. *Unwanted Stopping;*
5. *Electrocution;*
6. *Collision with object;*
7. *Possible collision with buffer-stops.*

*The top event hazard affected by this issue is 'Possible collision with another train' due to excessive RM driving].*

Step 2: Review of existing restrictions and identification of any new mitigation options against safety causes/issues; direct causes first and secondary causes after.

C. Risk assessment and selection of available options:

The aim of this part of the review is to assess the risk of identified causes/issues, effectiveness, and practicability of identified mitigations/ restrictions; and based on such an assessment, select the mitigations/ restrictions to be implemented. This is carried out in three steps:

Step 1: Review the causes/hazards before the restriction is implemented, applying the criteria P2: Is the level of the risk arising from the hazard contrasted against existing train/signalling system "Better"; "Worse" or "Comparable"?

Step 2: Review the need for mitigation and where needed the effectiveness of identified restrictions applying the criteria P2.

Hazard is considered mitigated to ALARP:

1. If the event is judged to be no more likely to occur without restriction or mitigation than equivalent events in the existing railway system, then no restriction is needed;

OR

If the proposed restriction or mitigation makes it no more likely that the hazard will lead to an accident than equivalent events in the existing railway system, then the restriction or mitigation is satisfactory,

AND

2. If no alternative or further restrictions or mitigations are needed to make the likelihood of an accident no greater than that from equivalent events in the existing railway system.

Step 3: Review the causes/hazards, assuming that the restrictions are implemented, applying the criteria P2: Is the level of the risk arising from the hazard contrasted against the existing train/signalling system "Better"; "Worse" or "Comparable"?

D. Review of identified mitigations and ALARP argument:

The aim of this final stage of the process is to rationalise the output of the workshop and agree the conclusions. An example of the output from the workshop is presented in a Table 7-9 below.

| ID | 1 |
|---|---|
| Details of the existing restriction | Temporary/Permanent Speed Restrictions (TSR/PSR) imposed by existing means (either maintaining 270 code or recommisioning coasting spots) will not be obeyed by 09TS. Maintenance briefing to reinforce existing rules to be published |
| Source | System design in overlay |
| Related Hazards | Possible derailment due to over speed |
| Cause - Description | Inappropriate maintenance action |
| Cause - Category | Primary |
| Mitigation Method | Maintenance and operations briefing to reinforce existing rules on the application of TSRs to be published. |
| Action | Publish maintenance and operations briefing to reinforce existing rules on the application of TSRs. |
| Actionee | Name Surname |
| Risk Assessment - Initial | Worse |
| Risk Assessment - Residua | Comparable |

**Table 7-9: Example of the restrictions analysis outcome**

### 7.6.3 Modelling

The Victoria Line Upgrade and Subsurface Lines Upgrade projects are introducing fundamental changes to the Railway infrastructure, including new trains and signalling. It is a legal requirement to demonstrate that these changes will not cause the risk of passenger fatality to increase beyond the current levels, and that the residual risk is ALARP and tolerable.

Both projects are being delivered through a series of migrations between defined Configuration States that map to key milestones in the programme, thereby changing the risk profile of the railway at each migration stage. To reflect this, the QRA model has been developed for each Configuration State, resulting in a set of risk predictions for passenger fatalities at each stage (defined and analysed in the Interim Safety Performance models).

To ensure that the total risk is kept to an acceptable level, as a minimum, a level which is no worse than at present, the QRA models are used to set safety targets for aspects and functions of the systems. As already mentioned, a set of QRA models developed within the LU was handed over to the projects to support the derivation of safety targets and the quantification of core safety requirements. After detailed examination of the LU model it was decided that it was not suitable for defining the current level of risk because:

1. A significant number of logical as well as data errors were identified;
2. The model logic was not fully representative of the current function and operation of the railway;
3. The base data was no longer traceable to a verifiable data source.

The models were reviewed and updated to provide a Current Safety Performance of the Victoria Line and of the Subsurface Lines (CSPVL and CSPSSL) QRAs. This revised CSPVL QRAs formed a baseline model of the current railway, and are being used to:

1. Confirm the current levels of risk, to set a baseline safety target;
2. Provide a baseline risk model that can be modified to reflect the changes to the railway, for each major migration stage.

Using the CSP models as a starting point, the Interim Safety Performance and the Final Safety Performance models have been developed to define the impact of changes to the railway on the system safety performance.

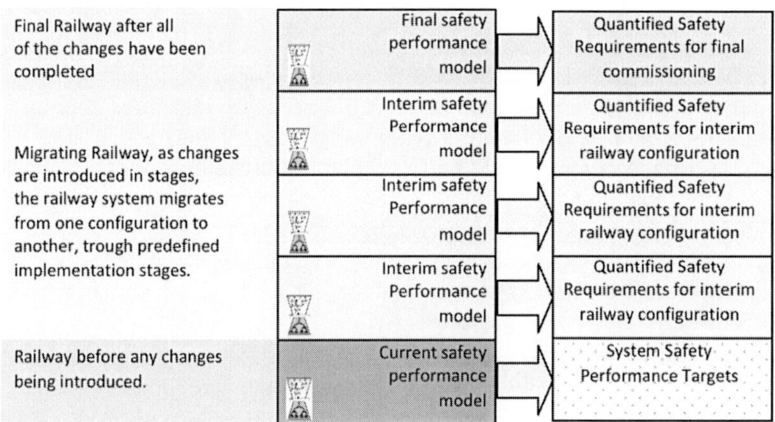


Figure 7-11: Development of QRA models in support of staged project implementation

To achieve this, the following activities were carried out:

1. Logic and data of models have been updated to include potential causes of minor and major injuries to all exposed parties (Passengers, Workers and Neighbours);
2. A new loss analysis, completed to predict the injuries and, subsequently, equivalent fatalities, was carried out in some cases;
3. Additionally, in future if the scope is altered, it may be necessary to develop further models to include all novel changes.

The ISP and FSP series of the QRA models are capable of:

1. Bringing together all the risks from the trains, signalling, track, stations and operators (for modelled hazards only);
2. Integrating with other QRA models (produced by suppliers e.g. Bombardier and Westinghouse) so as to allow for reuse of the data already produced and validated by suppliers;
3. Producing a risk profile given the expected passenger fatality rates resulting from each hazard;
4. Producing a single figure prediction of the risk of a fatality to a passenger or worker or neighbour, expressed as the risk to the most exposed individual;
5. Calculating the relative contribution (apportionment) of individual causes (interface/subsystem/procedural failures) to the overall risk, originating from each individual hazard.

A regular cross check between the content of the hazard log and the QRA is essential, for each system configuration, to ensure the completeness of both sets of analysis during the programmes' lifecycle. In order to track the changes between the different incarnations of the QRA models, a change control process has been followed to ensure that each change is transparent and traceable. This process generates a spreadsheet which lists the changes between the each version of the model for each of the model's element. This change control register is maintained as a separate spreadsheet for each model, and will be updated as new models are produced for each of the project stages.

## Scope of the QRA work

The LU 2003 QRA consists of 20 models of the top contributors to safety risk. These are:

1. Arcing;
2. Collision Between trains;
3. Collision Hazard (with Object);
4. Derailment;
5. Platform Train Interface- platform;
6. Platform Train Interface- train;
7. Station Area Accident (SAA)- escalator;
8. SAA- lift;
9. SAA- other;
10. Train fire;
11. Station fire;
12. Tunnel fire;
13. Lift fire;
14. Escalator fire;
15. Unauthorised access;
16. Explosion;
17. Flood;
18. Structural failure;
19. Power;
20 Ventilation.

The changes to the railway as a result of the VLU and SSL do not affect all of the models. The models can be split into two groups: models affected by changes, and those that are not.

| Model | Affected by scope of changes | |
|---|---|---|
| | VLU | SSL |
| Arcing | ✓ | ✓ |
| Collision between trains | ✓ | ✓ |
| Collision with object | ✓ | ✓ |
| Derailment | ✓ | ✓ |
| Passenger Train Interface on train | ✓ | ✓ |
| Passenger Train Interface on platform | ✓ | ✓ |
| Train fire | ✓ | ✓ |
| Tunnel fire | ✓ | ✓ |
| Station fire | ✓ | ✓ |
| Unauthorised access | ✓ | ✓ |
| Ventilation | ✓ | ✓ |
| Power | | ✓ |
| Explosion | | |
| Station Area Accidents - escalator | | |
| Station Area Accidents - lift | | |
| Station Area Accidents - other | | |
| Flood | | |
| Escalator fire | | |
| Structural failure | | ✓ |
| Lift fire | | |

**Table 7-10: LU 2003 QRAs affected by changes to the railway introduced by VLU and SSL**

Only the models affected by a scope of change have been subjected to a thorough review and update.

In order to track the changes between the LU 2003 QRA model and the CSP models, a change control process has been followed in order to ensure that each change is transparent and traceable. This process generates a spreadsheet which lists the changes between the two models for each model element.

These changes are of the following types:

Deleted – The model element has been deleted. This is generally due to a structural change, and the change rationale will briefly describe the justification for this change. Full details of the justification for the change are provided;

Added - The model element has been added. This is generally due to a structural change and the change rationale will briefly describe the justification for this change. Full details of the justification for the change are provided;

Data model or data change – This indicates a change to either the type of data model used and/or the data value used in the model. The rationale for this is described in the appropriate field of the modelling tool;

Loss data – This shows where loss data has been revised.

The Data Sources and References in the LU 2003 QRA model were generally not provided, whereas in the CSP model, a Data Source and Reference has been provided for each event. This change control register is maintained as a separate spreadsheet for each model, and is being updated as new models are produced for each of the stages. The QRA works in support of the Victoria Line Upgrade have been used in this report as an illustration of works done for both projects.

One of the main reasons for changes is that the Victoria line has some unique features that were not correctly reflected in the LU model:

1. It is all subsurface; consequently there are no embankments, bridges or external environmental effects;
2. It is all twin tunnels; consequently a train cannot collide with a train on the adjacent track except at crossovers;
3. It is all deep tunnel that is mainly straight;
4. There are no curved platforms;
5. There are no speed restrictions due to track curvature, even the points are arranged to be capable of full line speed;
6. The trains are driven automatically by the ATO;
7. The trains are protected by the Safety System; this is a protection system that is completely independent from the ATO train control system;
8. Due to the safety system there is no trip cock protection on the railway. A train driven in SM can pass a signal in danger without being tripped. On an automatic signal, a driver can proceed under rule without authority of the signaller, although he should inform the signaller.

The train can be driven in three modes:

1. Automatic using the ATO called Auto;
2. Manually with the safety system still providing protection called Coded Manual (CM) (in the event of ATO failure). Speed should be limited to 22mph, but is not enforced;

3. Manually at slow speed with no safety system protection called Slow Manual (SM) (in the event of safety system failure). Speed is limited to 10mph and is enforced by the train speed governor.

In addition to above the following issues were identified:

1. On all models large numbers of logic gates with no connection to a top gate or an event tree were identified;
2. A number of logical inconsistencies were identified and corrected;
3. None of the data used in the original models was substantiated.

The process of review and acceptance of the revised model logic was as follows:

1. The starting point was LU VL 2003 QRA;
2. The next stage was to identify the known specific characteristics of the VL;
3. The model was then modified to reflect these characteristics. As the model is VL specific, if a base was not applicable it would be deleted rather than having the data changed in order to result in no effect;
4. In order to complete the review process, the project organised a final series of workshops; and where appropriate, experts reviewed and accepted the logic as representative of the current VL;
5. A workshop report was issued to substantiate the changes proposed to the models;
6. Subsequent to the workshop report it was identified during the data analysis that there was a need for additional changes to the logic;
7. The additional changes were ratified at a review meeting with the VL Design Authority and Systems Engineering and LU Operations.

An identical process was followed on SSL for the review of the relevant QRA models. The principle change made was to merge the three QRA models for the SSR lines (the Metropolitan, District, Circle, and Hammersmith & City lines) into one model. This was done by recognising that the lines share the same infrastructure for a portion of their length, and subsequently simplifying the representation of the upgrade.

### *Summary of Data Update - Methodology*

The principle steps for validation of the CSPVL models, representing current configuration of Victoria Line, are listed below:

1. Set review scope;
2. Compile list of potential data sources, including scope and coverage;
3. Assess current data set:
    a. Compile set of new data items ("Obtain") and identify possible data source;
    b. Group data items into common sources;
4. Get data for each new data event;
    a. Define appropriate data model type;
    b. Review source;
    c. Assess uncertainty issues;
    d. Obtain data item(s);
    e. Write rationale, reference and assumptions.

***Summary of Data Update - Scope of the Review***

The aspects of the model event data which were reviewed are:

1. Local or Generic Data Models;
2. Data Model Type;
3. Rationale;
4. Reference.

***Summary of Data Update - Data Sources***

The data types used in the model were identified in the course of the model data review and are as follows:

1. INCA: INCA is the Incident Capture and Analysis database used by LU to capture incidents which occur on the Underground infrastructure. Data has been entered into this database since 1992. Incident data is entered under a particular classification, of which there are 105 different categories. For the data items used in the model, the 14 year period between 1992 and 1996 has been used. The aim has been to use incident data from the Victoria Line only. However, in the case that there were few or no incidents of a particular type on the Victoria Line, then the search has been extended to then the Bakerloo-Central-Victoria system and, failing that, then the whole Underground system. Where this has been done, the results have been reviewed to ensure applicability.
2. RAM: A number of the event data items in the model are due to equipment failure rates. The data for these items is common with that used for system availability modelling in the VLU project.

Data on these failures in use is captured in the CuPID database, and it is from this database that the data has been derived.

3. Asset Data: There are a small number of asset data items which are needed in the model. However, it should be realised that all of the incident data contained in both the INCA and CuPID database is from a system which has a particular set of physical and operational conditions. It is an assumption in the use of these databases that the conditions over the time period that the data was obtained are sufficiently similar to the current system to allow the use of the data.
4. Passenger Occupancy Data: There are two types of passenger occupancy data which are used in the model. The main aspect of passenger occupancy which is required in the event trees, is the division between high, medium and low passenger loadings.
5. Human Factors: There are a large number of events in the model which are human errors. The probability data for human errors regarding skill and rule-based tasks is generally in the range 0.01 to 0.001. While the data provided for human errors is largely presented without justification, it is within this range.
6. Timetable/Number of trains: One of the operational characteristics which affect the risk is the number of trains running on the system. This number has been taken from the signalling availability model and it is one of the operational parameters which describe the system.
7. Judgement: A large number of the failure events in the model are estimates based upon judgement. Where this is the case, operational or engineering expertise has been sought from competent personnel.

The QRA data set has been assessed against the aspects listed above. As part of this, the most appropriate data source for each data item has been identified. The data sources for the different models are shown in Table 7-11below.

This table lists the number of events for each of the data sources listed above. It can be seen from this table that, in addition to the set of data sources listed above, there are two additional categories: "Model" and "Delete". The designation "Model" is used for those elements which are identical to an element in another part of the model, or in another model. In this case, the rationale is duplicated, and reference is made to the other model element. The designation "Delete" is made for model elements which are to be deleted. This can be for a number of reasons, as follows:

1. The model element is not used and has no dependencies;
2. The model element is the complement of another model element;
3. The model element is the duplicate of another model element.

Where a model element has been designated "Delete", the rationale for this has been given.

| Model Name | INCA | Judgement | Passenger Occupancy Data | HF | RAM | Clarify | Time table | Asset Data | Model | Delete |
|---|---|---|---|---|---|---|---|---|---|---|
| Arcing | 20 | 12 | 0 | 5 | 3 | 0 | 0 | 3 | 45 | 0 |
| Collision between trains | 0 | 28 | 0 | 22 | 18 | 2 | 3 | 4 | 48 | 169 |
| Collision with object | 7 | 0 | 3 | 7 | 10 | 0 | 0 | 2 | 50 | 7 |
| Derailment | 4 | 67 | 0 | 12 | 8 | 1 | 1 | 1 | 107 | 165 |
| PTI on train | 7 | 13 | 0 | 7 | 6 | 0 | 0 | 0 | 1 | 185 |
| PTI Platform | 17 | 37 | 0 | 3 | 2 | 0 | 0 | 0 | 1 | 2 |
| Train fire | 10 | 19 | 8 | 0 | 0 | 0 | 0 | 0 | 5 | 20 |
| Tunnel fire | 10 | 13 | 3 | 24 | 13 | 0 | 0 | 0 | 0 | 9 |
| Station fire | 25 | 20 | 4 | 20 | 14 | 0 | 0 | 2 | 0 | 24 |
| Unauthorised access | 3 | 9 | 0 | 0 | 0 | 2 | 0 | 0 | 0 | 3 |

**Table 7-11: Model Element Data Sources Overview**

### *Principles for Data Validation in the QRA Model*

The development of a model structure is intrinsically linked to the data available to support the model. In general, the level at which the data is available is the level of detail to which the model should be developed.

Sometimes this is in conflict with other model requirements. For example, if there is a need to use the model to assess the effect of changes which are at a higher level of detail than the data supports, then there may be a justification for applying subjective judgement techniques to estimate parameters.

In the QRA model, if the model structure is not supported by the data, then an assessment needs to be made as to whether the elements of the structure are required for analysis of system changes. If they are not, then the structure should be simplified to the level that the data supports.

***Loss data***

The loss data for each consequence was checked for validity both in the context of the specific model, and in the context of the model as a whole. The basis for this checking was other industry models such as the Railway Standards and Safety Board Safety Risk Model and the West Coast Main Line Current Safety Performance Model.

However, it should be noted that as these models are for heavy rail only; the loss data could not be directly compared in many cases and historical data was used as a reference point, including a set of incident data for train accidents in tunnels.

The overall approach was to leave the loss data unchanged from the LU 2003 QRA model unless a specific discrepancy was noted.

The loss data will be further reviewed if affected by changes in the VLU or SSL.

***Summary of the Initial Findings***

The first finding of the data review was that information regarding the data (the rationale) was not available in a single source.

Instead, descriptions to different levels of detail were contained in both the FT+ model and in a number of other reports. Hence, the first task was to gather all of the information into a single source so that it could be reviewed. This was done by exporting all of the FT+ model events as Excel spreadsheets, and importing into these all of the descriptions from the other reports.

These spreadsheets formed the basis for the model review with which reference to the model structures in FT+ where required. It is desirable that all information regarding the event data (excluding the change history) should be held in the FT+ model. However, there are restrictions in terms of the number of notes fields available (8) and the size of those fields (255 characters). Also, some of these fields may be required for description and assumptions regarding the logic.

Hence the available fields have been used as follows:

1. Reference: Data source reference including any spatial or temporal aspects;
2. Data source: A description of how the data items were derived from the data source;

3. Planned Maintenance;
4. Assumptions.

Where additional information was required, which did not fit in these 4 fields, then reference has been made as to where it can be found either in the Data Gathering and Validation report, or the data change register/database which exists outside of the model.

***Results***

Having validated the CSPVL model data set, the next task was then to import all of the data into the FT+ models so that the models could be run and the overall results checked against the existing LU models. An example of the results for the Collision with Trains model is presented and discussed below. The LU model divides collision between trains into Fast, Medium or Slow and Head on Fast or Slow. The causes are mainly attributed to signalling wrong side failures or the driver passing a signal at danger in SM or reversing.

The major changes of the QRA logic, in the CSPVL model, were:

1. That 'Failures in Coded Manual mode' are added. This is specific feature of the Victoria line signalling ;
2. The 'interlocking sets' conflicting routes' are added. This base event was missing from the original QRA model;
3. The 'train driven in SM following safety system failure' is added. This is a specific feature of the Victoria line signalling;
4. The modelling of 'Automatic Train Operation related failure' is improved.

In terms of data, the main changes made are as follows:

1. The LU 2003 QRA passenger train loadings have been reviewed and revised. In the LU 2003 QRA model, High loading was estimated to occur 0.1% of the time (i.e. one minute of each operational day), medium loading 2% of the time and low loading 97.9% of the time. These have been revised to: high loading 10%, medium loading 16%, low loading 74%;
2. The LU 2003 QRA model assumed complete segregation of tracks, whereas the CSPVL model has taken into account the two crossovers on the VL.

The following are Risk Results from the updated CSPVL Collision with Trains model.

| CSPVL | LU 2003 QRA |
|---|---|
| 0.067 | 0.062 |

**Table 7-12: Total Risk (fatalities/yr)**

| Original Model LU 2003 QRA Model | | Updated CSPVL Model | |
|---|---|---|---|
| Event Description | Frequency | Event Description | Frequency |
| Fast collision between trains | 1.39E-02 | Fast collision between trains | 1.23E-04 |
| Head on collision between trains | 0.00E+00 | Head on collision between trains | 3.52E-02 |
| Slow speed head on collision between trains | 4.18E-04 | | |
| Medium speed collision between trains | 1.83E-03 | Medium Speed collision | 4.43E-03 |
| Slow speed collision between trains | 3.08E-03 | Slow speed collision of two trains | 4.32E-08 |
| Total | 1.92E-02 | Total | 3.98E-02 |

**Table 7-13: Comparison of Initiating event frequencies**

At the level of initiating the event, the CSPVL model calculates the frequency to be approximately 50% less than that calculated by the LU 2003 QRA model. Given the rarity of this type of event, this degree of variation is not significant.

In the event tree, the risk calculated by the CSPVL model is slightly greater than that calculated by the LU 2003 QRA model.

While there is a small effect from the inclusion of head on and subsequent collisions in the CSPVL model, this difference is almost entirely due to the different apportionment to passenger loading levels, which has been explained and justified above.

***Conclusions from the review of the baseline QRA and definition of Tolerable Hazards' Rates***

A summary of the risk results for all of the 10 models updated for the Victoria Line is presented below.

| Model Name | CSPVL Risk | % of VL Total | LU VL 2003 QRA Risk | % of LU VL 2003 QRA Total | % change between CSPVL and LU VL 2003 QRA |
|---|---|---|---|---|---|
| Arcing | 1.86E-03 | 0.30% | 2.09E-03 | 0.22% | -11% |
| Collision between trains | 0.068 | 10.86% | 6.19E-02 | 6.46% | 10% |
| Collision with object | 3.25E-03 | 0.52% | 1.90E-03 | 0.20% | 71% |
| Derailment | 0.0388 | 6.20% | 5.95E-02 | 6.21% | -35% |
| PTI on train | 0.138 | 22.04% | 4.78E-01 | 49.91% | -71% |
| PTI/platform | 0.331 | 52.86% | 2.25E-01 | 23.49% | 47% |
| Train fire | 0.0138 | 2.20% | 8.48E-02 | 8.85% | -84% |
| Tunnel fire | 2.10E-05 | 0.00% | 1.12E-03 | 0.12% | -98% |
| Station fire | 4.58E-03 | 0.73% | 9.54E-05 | 0.01% | 4701% |
| Unauthorised access | 0.0269 | 4.30% | 0.0433 | 4.52% | -38% |
| Total | 6.26E-01 | | 9.58E-01 | | -35% |

**Table 7-14: Summary of Risk results**

It can be seen that there is a decrease in risk of 35% over the LU 2003 QRA total for these 10 models. There have been many changes between the two sets of models, both structurally and in terms of data. The principle reasons for the differences are as follows:

1. A number of changes in model structure, data and loss apportionment have contributed to differences in the risk in the CSPVL Derailment, Train Collision and Collision models as compared to the LU 2003 QRA models;
2. The risk due to PTI on Train has reduced by 71% in the CSPVL model, as a result of an improved alignment with INCA data. As this model constituted 50% of the LU 2003 QRA total risk for these 10 models, this reduction is the main contributor to the decrease in the total.
3. All of the fire models have changed significantly, though they are not major contributors to the overall risk. This is due to changes in model structure that enable the models to be closely based upon incident data.

All of the above comparisons have been made against the LU 2003 QRA as, due to the low incidence of fatality on the LU network as a whole (averaging 4 per annum, excluding suicide), it is not possible to conduct comparisons at any level of detail or accuracy.

This is consistent with the low frequencies of the initiating events which are in general of the order of one in one hundred or less. It should be remembered that the aim of this validation exercise is to provide a baseline set of models which is representative of the current Victoria Line on which the operational and infrastructure changes of the different stages of the VLU can be imposed and effects on safety assessed.

On this basis, the conclusion was reached that this updated set of the CSP QRA models is a suitable representation of the hazards on the VL. As such, it provides a suitable baseline which was used to set a baseline safety target, Tolerable Hazard Rates (THRs). From that, the total contribution to overall risk, for each Core Hazard, from failures of the new signalling equipment, was calculated. Only the apportionment to the new signalling system, and to the rolling stock, was calculated, since other changes to the Victoria Line are conventional changes, and no new or novel equipment was being introduced.

| Top Event and Risk (fat./yr) THR | | Apportionment to Signalling (contribution) and Notes | |
|---|---|---|---|
| Arcing | 1.86E-03 | 0.00% | |
| Collision between trains | 6.81E-02 | 1.28% | A safety factor of 2x has been applied to account for signalling sub-system failures which may not have been modelled in detail in the CSPVL (Baseline) model. |
| Collision with object | 3.25E-03 | 0.00% | Signalling contribution is non-zero but negligible. |
| Derailment | 3.88E-02 | 4.45% | |
| PTI on train | 1.38E-01 | 0.00% | |
| PTI platform | 3.31E-01 | 0.00% | |
| Train fire | 1.38E-02 | 0.00% | |
| Tunnel fire | 2.10E-05 | 0.00% | |
| Station fire | 4.58E-03 | 0.00% | |
| Unauthorised access | 2.69E-02 | 0.00% | |

**Table 7-15: Victoria line THRs**

The above data (apportionment) were calculated by identifying all of the events in the models relevant to signalling, and through conducting an analysis of the cut set to calculate the contribution to the overall individual risk for that Core Hazard.

The aim of this analysis was to allow the supplier of the signalling system to consider the cost, for each top event, of an "effort" to reduce the risk for an individual hazard from (say) 1E-9/hr to 1E-10/hr; and if the predicted costs are comparatively small, to justify a general statement, that further cost benefit analysis is not necessary for each top event.

The ALARP argument was produced (in line with LU Standard 2-05101-101) using a value of £1.4 million as the value of avoiding a fatality (VAF), to calculate the safety benefit associated with a measure. A factor of 3 was applied to the safety benefit to assess whether the cost of a measure is grossly disproportionate to the achieved safety benefit.

However, this did not preclude any analysis (quantitative or qualitative) of specific issues where required.

### *Development of Interim Safety Performance Models*

The overall railway level safety requirements are that:

1. The levels of risk after the change shall be ALARP and tolerable, and
2. No greater than the current level of risk.

For the VLU project, the update of the QRA that was reflecting the system changes implemented as part of the first system configuration (referred to as V2.1), has been used in support of the safety argument presented in the System Safety Case (elaborated later in this book). This QRA model is referred to as ISP V2.1 (Interim Safety Performance for migration stage/system configuration V2.1). This update included a review of the VL QRA model against the models developed by the rolling stock and signalling system suppliers in order to ensure that the models are sufficiently consistent to test if the tolerable hazard rates and other targets used by the suppliers are suitable to reduce risk to an acceptable level. The ISP V2.1 QRA model has been used to assess the anticipated level of risk at the railway system configuration compared to the baseline level as estimated by the CSPVL Baseline model.

As part of the system configuration V2.1, an extra pathway in off-peak hours (after 20:30) on which to test the new system, was created (the new train did not carry any passengers). Hence the modelling approach was to change the baseline model to represent the risk posed by the new train on this single train path, the total risk is the risk from the baseline model (which is unchanged as V2.1 does not alter the existing system nor the existing service) together with the risk from the new ISP V2.1 model (representing the risk from a new, overlaid system and the additional train path). It should be noted that as the QRA models the risks to passengers only, rather than the effect on the risk to the drivers, testers or maintainers, these latter considerations have not been addressed in these models. These risks have been assessed separately using the comparative methodology described earlier. As a result of this approach, for each configuration of the system architecture for which a System Safety Case is being prepared, a representative QRA model will consist of a Baseline model with altered data to reflect the changes, and the model reflecting the logic and data changes provided by improvements or new system components such as new trains. This approach results in there being two risk models to model the overall risk of V2.1 operation: one which models the existing system and one which models the new system. This allows each model to represent the different functionality of the old and new system and any interfaces between the two to be explicitly represented.

As more new trains are introduced to replace the existing rolling stock, the initiator data in each model is changed accordingly so that the risk from each system can be calculated at each migration stage. When all of the old rolling stock is replaced by the new train, then the existing system risk will be zero (as the initiator data will be zero) and all of the risk will reside with the new system. This can be represented by the following formula:

$$R_{totalVx}^{ISP} = R_{Vx}^{CSP} + R_{newVx}^{ISP}$$

Where

$R_{totalVx}^{ISP}$ Total risk for migration stage Vx (fatalities/yr);
$R_{Vx}^{CSP}$ Risk from operating with existing signalling and rolling stock for migration stage Vx (fatalities/yr);
$R_{newVx}^{ISP}$ Risk from operating with new signalling and rolling stock for migration stage Vx (fatalities/yr).

The first task, in order to represent ISP V2.1 in the upgrade of the CSPVL models, was to review each of the ten hazard models relevant to the VLU to assess if and how these are affected by the changes implemented in phase V2.1. The following Table 1 presents the results of this assessment.

| Top event | Effect on passenger risk for V2.1 new train service | Changes to risk model |
|---|---|---|
| (potential for) Collision with trains | As it is assumed that the risk from collision between 67TS remains unchanged, the new risk is that a failure of the T2 train control and protection causes a collision with a 67TS resulting in passenger fatalities on the 67TS train.<br>It is assumed that the new signalling system will not negatively affect the wrong side failure rate of the existing signalling system. | The baseline train collision model has been altered to represent the risk of collision due to failure of the new signalling and/or train.<br>All initiator data has been changed to represent the operation of a single train path for 4 hours per day.<br>Exposure data in terms of train loading (of the second as the T2 is not carrying any passengers) has been modified to represent operation in off-peak hours only.<br>These changes result in a new $ISP_{V2.1}$ model. |
| (potential for) Collision with Object | The risk from the Collision Hazard for the 67TS remains unchanged. The only new risk is that the object derails the train and subsequent to derailment the T2 train is hit by a train carrying passengers | For T2, the consequence analysis has been changed to remove fatalities resulting from Derailment with no subsequent collision and where subsequent collisions occur to only include the effects of loading of the 67TS ( rather than combined loading).<br>The causes for objects on the track remain unchanged.<br>All initiator data has been changed to represent the operation of a single train path for 4 hours per day.<br>Exposure data in terms of train loading (for a subsequent collision, as the T2 is not carrying any passengers) has been modified to represent operation in off-peak hours only.<br>These changes result in a new $ISP_{V2.1}$ model. |

| Top event | Effect on passenger risk for V2.1 new train service | Changes to risk model |
|---|---|---|
| (potential for) Derailment | The only new risk is that T2 derails and is subsequently hit by a 67TS. | The 67TS risks remain unchanged. The baseline model has been developed to exclude fatalities from derailment without subsequent collision and in the event of subsequent collision only includes the 67TS loading. The causes of derailment (the fault trees) have been changed to take account of the new signalling and trains. These changes result in a new $ISP_{V2.1}$ model. |

**Table 7-16: Sample of QRA changes relevant to ISP**

Having identified where changes to the risk models were required, a more detailed review of each model was conducted to identify specific gates and base events concerning aspects of rolling stock and signalling for which a mapping to the rolling stock and signalling fault trees was sought. The V2.1 risk assessment has modified the Victoria Line baseline QRA models to take account of the following:

1. The train brake and 'traction and control' failure rates as defined in the supplier's model (BT VLU Vehicle level Fault Trees), have been used in the $ISP_{V2.1}$ model;
2. The DTG-R signalling system failure rates as defined in the supplier's model (WRSL Signalling and Train Control System Preliminary Fault Tree Hazard Analysis), have been used in the $ISP_{V2.1}$ model;
3. The supplier's model (BT VLU Vehicle level Fault Trees) did not include failure rates for the mechanical or structural elements, e.g. bogies, wheels, suspension etc. Where rates were included in the model, the figures from the baseline were used on the grounds that the design was based upon standards and best practice, and so the rates should be similar to those achieved on the 67TS;
4. The use of Connect digital radio to replace the cab radio and allow the train operator to communicate directly with other operators as required is included in the $ISP_{V2.1}$ models;
5. The V2.1 risk analysis assumes that the T2 train is used for a year (all rates are per year and all Initiators are events per year). The

total risk has been scaled for the relative proportion of usage of T2 and 67TS trains in each of the different phases. In V2.1 only one train path in off peak hours (after 20:30) was created for the T2 train;

6. The V2.1 loss analysis has been adjusted for the fact that one of the trains has no passengers. Also, T2 collisions can only occur at off peak low loading and all losses associated with high or medium loading have been set to zero;
7. The V2.1 models, model the risk from the T2 train operating without passengers on the extra path which has been created in off-peak hours. The risk from this operation had been added to that from the Baseline model to obtain the total risk from the V2.1 operation. Hence, the data changes were of two types: the changed failure data for the new rolling stock and signalling systems and the changed exposure data in terms of number of train movements;
8. The changed data for the new rolling stock and signalling systems were obtained from the suppliers fault tree models. In some cases these values were set as requirements which the systems must be shown to meet. It is assumed that the supplier's assurance cases will demonstrate how these targets or requirements will be met, and testing in engineering hours will highlight any potential issues. Any requirements or targets not achieved had to be resolved before the new train goes into service in passenger hours.

The assessment of the impact of system changes on its safety performance required the use of ISP QRA

A summary of the changes made to the models to represent V2.1 and a comparison of the $R^{ISP}_{totalV2.1}$ risk against the CSPVL baseline risk is presented below.

As the change in V2.1 was to introduce a new train path, an increase in risk over the baseline is expected. However, it can be seen that this increase is negligible or zero for the 10 hazards affected by the VLU. The results of this analysis have been used in support of the safety argument within the safety case, as detailed later.

| Model Name | $R^{CSP}_{Vx}$ | $R^{ISP}_{totalV2.1}$ | % change from CSPVL Baseline |
|---|---|---|---|
| Arcing | 1.860E-03 | 1.860E-03 | 0.000% |
| Collision between trains | 6.800E-02 | 6.800E-02 | 0.001% |

| Model Name | $R_{Vx}^{CSP}$ | $R_{totalV2.1}^{ISP}$ | % change from CSPVL Baseline |
|---|---|---|---|
| Collision with object | 3.250E-03 | 3.266E-03 | 0.490% |
| Derailment | 3.880E-02 | 3.882E-02 | 0.054% |
| PTI on train | 1.380E-01 | 1.380E-01 | 0.000% |
| PTI platform | 3.310E-01 | 3.310E-01 | 0.000% |
| Train fire | 1.380E-02 | 1.380E-02 | 0.000% |
| Tunnel fire | 2.100E-05 | 2.100E-05 | 0.000% |
| Station fire | 3.860E-03 | 3.870E-03 | 0.258% |
| Unauthorised access | 2.69E-02 | 2.693E-02 | 0.100% |
| Total | 6.255E-01 | 6.255E-01 | |

**Table 7-17: Summary of Risk results**

***Apportionment and confirmation of safety targets***

As already elaborated, the baseline QRA model (CSPVL and CSPSSL) defines the Tolerable Hazard Rates (THR) for the Victoria Line and Subsurface Lines railway systems, for each of the core hazards. In order to apportion the THRs to specific constituents of the system, for which a quantified target (or a failure rate) is required in support of the safety requirements' quantification, further analysis has been done as follows:

1. The system elements affected by the changes brought about through implementation of VLU and SSL project scopes have been identified and the changes classified as per each subsystem (signalling, rolling stock, power, etc);
2. The impact of these changes on the QRA model was analysed and scoped;
3. The QRA elements have been classified to enable apportionment;
4. The contributions to the overall risk for each of the core hazards has been calculated.

For example, for the Victoria Line, and the results of this analysis of the signalling contribution to risk, is calculated from the CSPVL QRA model, are shown in the Table 7-18 below.

| Top Event | Risk (fat./yr) | Signalling Contribution | Notes |
|---|---|---|---|
| Arcing | 1.86E-03 | 0.00% | |

| Top Event | Risk (fat./yr) | Signalling Contribution | Notes |
|---|---|---|---|
| Collision between trains | 6.81E-02 | 1.28% | A safety factor of 2x has been applied to account for signalling sub-system failures which may not have been modelled in detail in the CSPVL model. |
| Collision with object | 3.25E-03 | 0.00% | Signalling contribution is non-zero but negligible. |
| Derailment | 3.88E-02 | 4.45% | |
| PTI on train | 1.38E-01 | 0.00% | |
| PTI platform | 3.31E-01 | 0.00% | Signalling contribution is non-zero but negligible. |
| Train fire | 1.38E-02 | 0.00% | |
| Tunnel fire | 2.10E-05 | 0.00% | |
| Station fire | 4.58E-03 | 0.00% | |
| Unauthorised access | 2.69E-02 | 0.00% | |

**Table 7-18: An analysis of the signalling contribution to risk as calculated from the CSPVL**

This data was calculated by identifying all of the events in the models relevant to signalling, and by conducting an analysis of the cut set to calculate the contribution to the overall individual risk for that hazard.

A similar exercise was completed in support of the Subsurface Lines Upgrade Programme, in particular in support of the quantification of the safety requirements intended for tendering a new signalling system. Further details and examples of this work are presented in the following section.

Specifically, on the Victoria Line project for signalling and rolling stock subsystems, the suppliers developed their own safety targets early in the life cycle of the project, prior to the development of the overall London Underground safety targets described earlier. Therefore, the only pragmatic solution was to use the developed ISPVL and FSPVL models in order to confirm that the solution being provided by the supply chain will deliver, as a minimum, safety performance commensurate with the safety performance of the existing subsystems (signalling and rolling stock).

This was done by the substitution of relevant parts of logic or data in the CSPVL model to derive the ISPVL, model and the results are presented in Table 7-17.

As a result of this work, the supply chain safety targets were accepted.

Furthermore, this information was used to support the supply chain's ALARP review, allowing consideration for how much it will cost for each top event to reduce the risk for an individual hazard from (say) 1E-9/hr to 1E-10/hr and if the predicted costs are disproportionate, to justify a general statement that a detailed cost benefit analysis is not necessary for each top event.

The ALARP argument was produced in line with LU Standard which requires that "a value of £1.4 million is used as the value of avoiding a fatality (VAF) to calculate the safety benefit associated with a measure".

A factor of 3 was applied to the safety benefit to assess whether the cost of a measure is grossly disproportionate to the safety benefit achieved.

## 7.7 Derivation of safety requirements

As already mentioned earlier, safety requirements can be grouped into two categories:

1. Requirements related to compliance with standards and good practice;
2. Specific system safety performance related requirements.

Requirements related to compliance with standards and good practice are relatively easy to identify, and on both SSL and VLUP these were identified by the safety team in the early stages of the lifecycle.

In the second category, specific system safety performance requirements were elicited through the review of mitigations/protection measures in the hazard log. Those mitigations/protection measures that are related to novel equipment, or new processes, or any novel environment states within which the conventional equipment or a process is to operate, were converted into safety requirements. Those safety requirements that are related to novel equipment were further analysed in order to facilitate the representation of these in the QRA, either as base events or logical gates.

Quantified targets were set based upon these requirements. Table 7-19 presents an example of the VLUP safety requirements, with 4 quantified requirements (R00001, 25, 26 & 28) in italic font.

| Requirement | Description |
|---|---|
| *R00001* | The CSPVL QRA shows that in order to meet the overall goal of achieving risk levels proportionate to those in the overall LU risk model the risk due to collision between trains on the Victoria Line should be not greater than $6.81 \cdot 10^{-2}$ fatalities per year. |
| *R00005* | All equipment specified shall be compliant with new EMI/EMC legislation. |
| *R00006* | All trainborne and trackside communications equipment shall not be susceptible to trainborne EMI/Regenerative braking. |
| *R00007* | Configuration control of Automatic Train Protection (ATP) map data shall be enforced throughout the life cycle of the system (Metronet asset performance issue). |
| *R00008* | Configuration control of trackside and train borne ATP data shall be implemented. |
| *R00009* | Configuration management of track capability data linked to signalling data shall be implemented. |
| *R00011* | Design of the track circuit configuration shall be compatible with 09TS (e.g. straddle existing short depot track circuits). |
| *R00012* | Door alarms to be fitted on Signalling Equipment Room (SER). |
| *R00019* | Mobile and portable radio equipment shall be banned from SERs (and other areas) if analysis shows this to be necessary. |
| *R00022* | Operational procedures for safety critical communication between the train operator and the controller shall be reviewed and updated and appropriate training provided. |
| *R00023* | Power and signalling cables shall be sufficiently separated electromagnetically to comply with Standards. |
| *R00026* | The likelihood that the Automatic Train Operation (ATO) system attempts to exceed the stopping point shall not be greater than 2.44E-5 to ensure the same level of safety performance as today. |

| Requirement | Description |
|---|---|
| *R00028* | The likelihood of the Emergency Brake being unavailable on two or more cars shall not be greater than 1.49E-7 to ensure the same level of safety performance as today. |
| *R00042* | The overlay of the new signalling system must not affect the operation of the existing signalling system |
| *R00046* | The track lubrication system shall be designed to ensure that relevant standards have been adhered to and that the performance of the railway targets can be met. |
| *R00047* | Trainborne EMI/Regenerative braking shall not degrade or incapacitate train borne or trackside communications equipment. |
| *R00049* | Validation of data shall be carried out for each change. |

**Table 7-19: Examples of Victoria Line related Safety Requirements**

## 7.8 Construction of the safety arguments logical network – Safety Justification & Case

In support of both projects, VLUP and SSL, the safety justifications and safety cases were produced (and are still being produced), as an outcome of the above described safety management process. As an illustration of this, this section will present one example of the development of the safety justification and a safety case from VLUP, and an example of the hierarchy of safety cases from SSL.

As already outlined, the process models inform the Change Safety Analysis (CSA). The output of the CSA feeds the hazard log and the identification of the safety requirements, as illustrated by Table 7-12 below.

This information is used as a base for further safety analysis, identification and assessment of effectiveness of the protection measures and safety requirements, and construction of the safety fact net.

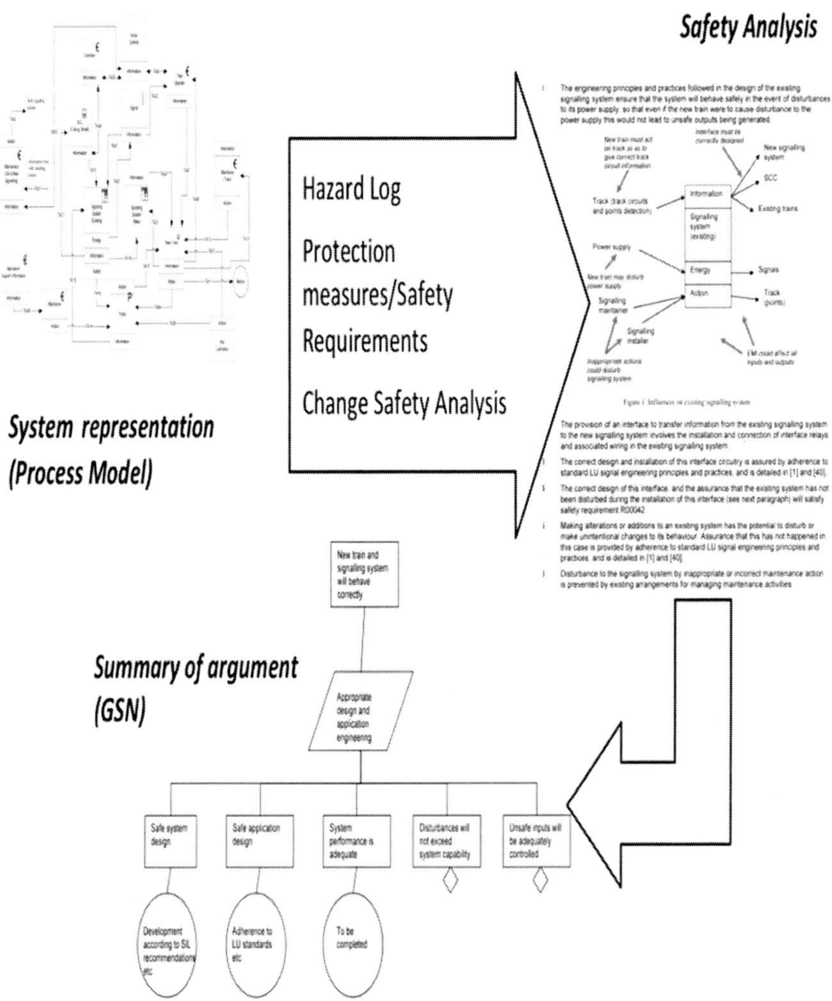


Figure 7-12: Development of Safety Arguments

### 7.8.1 Safety Justification

Usually, testing is the first life cycle stage that calls for a safety argument to be produced in support of carrying the testing or a trial operation. Production of such a "Safety Justification" is the first step towards the production of the safety case. Most often, the process for the derivation of the "Safety Justification" is exactly the same as the process for the production of the safety case. However, the overall delivered safety performance of the changed system, to enable the testing, is most often less onerous than in the case of the system being fully commissioned. This is due to the fact that the testing is done in a controlled environment, and that it is possible to rigorously enforce many operational controls that must be in place in order to compensate for the unproven or still unsatisfactory safety performance of the system.

The safety justification is based around the proof of the adequate management of hazards, restrictions and incident control and management, and an argument that these three key elements of the safety management system deliver safe system for testing. As an example from VLUP, a safety justification that was produced in support of the testing of the new train and the new signalling system in non-traffic hours is presented on Figure 7-13 below.

This particular testing was carried out under the operational control of London Underground, whilst the train-signalling system was delivered by the supply chain, including the proof of safety of the technical solution. The author structured the system safety justification around the three lines of reasoning, mentioned above, and an additional justification of the technical safety performance of the subsystem delivered by the supply chain, as follows (the structure of the safety justification is self explanatory (Figure 7-13):

1. Management of restrictions is adequate;
2. Applicable hazards are managed to closure or are conditionally closed and managed through restrictions. The Safety risk is ALARP and tolerable;
3. Incident management is appropriate;
4. Train-signalling system is acceptably safe subject to adequate management of residual risk through restrictions.

The conclusion of the Safety justification was that a set of restrictions, dependencies and other safety measures had been identified which enabled the train operations, associated with V2.01 testing, to be performed at a level of risk which is tolerable and ALARP. This Safety justification can thus be seen a subset of the reasoning model that is the System Safety Case for VLUP.

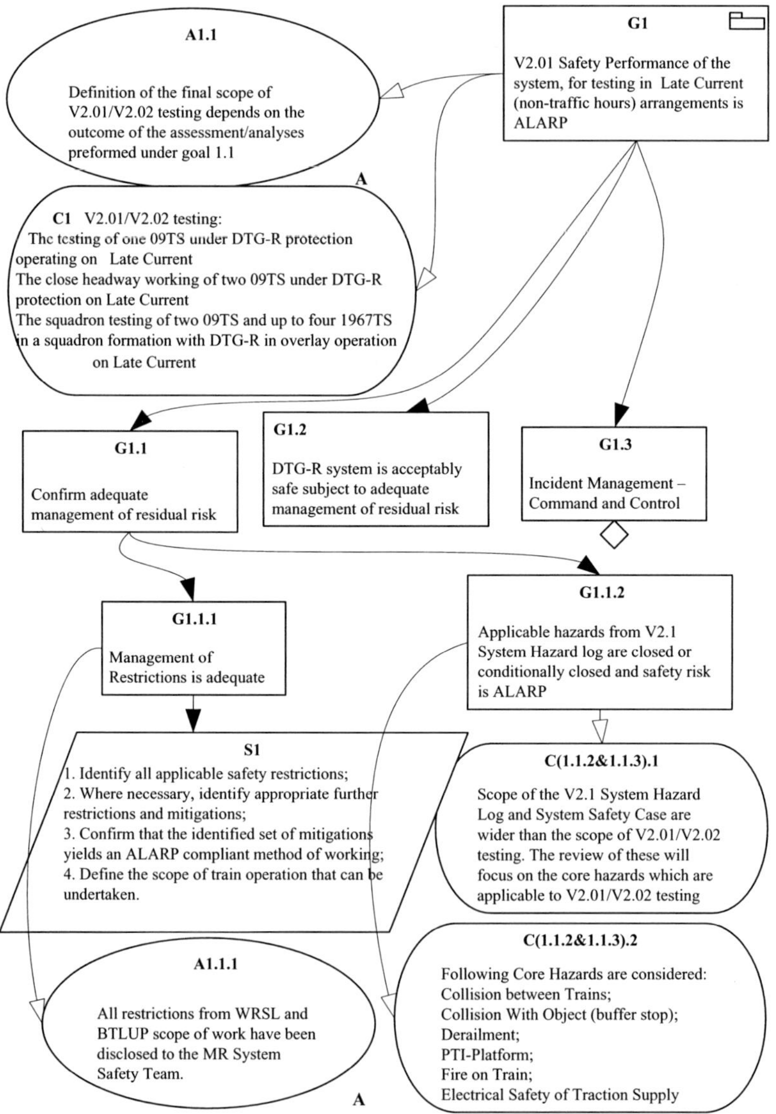


Figure 7-13 : Example of Safety Justification produced in support of the VLUP Testing of the new Train and Signalling system (DTG-R signalling system in overlay)

## 7.8.2 Safety Case

In general, the safety argument about any undertaking should be based around the following lines of reasoning, for each of the 4 elements (as discussed earlier):

1. Objectives/Requirements:
   a. Hazards, emerging from the change to the system, are identified, analysed and managed;
   b. Suitable safety targets are identified, selected and achieved;
   c. Safety performance of the system is commensurate with the ALARP principle;
2 Arguments including:
   a. Completeness of the analysis and implementation of protection and prevention measures (mitigations, barriers and containment measures), requirements and targets;
   b. Control of hazards on an individual basis including Qualitative Safety Arguments and/or Quantitative Safety Arguments;
   c. Options and Impact analysis (is there anything else, practicable, that can be done to reduce the risk level further) including the practicability argument (qualitative or quantitative cost-benefit analysis), assessment and comparison of the costs of any proposed changes to the system against the reduction in risk levels with the acceptance/rejection criteria for selection of options;
   d. Rationale for the adequacy of safety management activities and implemented mitigations, requirements and safety targets;
   e. Rationale for the adequacy of evidence provided in support of the arguments;
3. Evidence of:
   a. A comprehensive change safety management process has been followed;
   b. A comprehensive processes used to identify, manage and close hazards including: HAZOP, HAZID, Functional Failure Analysis, FTA, Interface Analysis, Single Point Failure Analysis and Diversity Analysis with traceable detailed records of these activities;
   c. The system design approach, including the design of the system architecture, use of a suitable means of checking for

errors, including design reviews and the ability of the system to self-check;

d. Applicable standards and practices being identified (such as EN50126, EN50128, EN50129), followed and complied with;

e. Use of a Problem/Fault/Data Reporting And Corrective Action System;

f. Work performed by the ISA (or Competent Independent Person) during the project;

g. Evidence of implemented protection and prevention measures (mitigations, barriers and containment measures), requirements and targets being achieved, including testing and verification and validation results;

h. Where novel process elements are employed, (e.g. by the introduction of new techniques not included in EN50126/8/9), a more extensive justification is required that they are consistent with the achievement of ALARP;

4. Context:

a. Outline of the system to be changed;

b. Outline of the changes to be introduced to the system with their impact on the system performance;

c. Outline of the operational environment of the system, including the outline of the technical "operational envelope" of the system operating within the given environment under given operating conditions.

### *The System Safety Case Structure*

On both, VLUP and SSL, the same philosophy for the construction of the safety case was adopted. An outline of the safety argument is broken down into:

1. Introduction: Safety objectives, targets and requirements are outlined in this section (summary of point "1" above);
2. System Definition: This section provides context for the safety case (point "4" above);
3. Quality and Safety Management Reports: This section presents the Quality and Safety Management System adhered to during the life cycle phases of the system development/change (point "3" from "a" to "f" above);
4. Technical Safety Reports: This section of the safety case contains all safety reports related to the system. This section corresponds to

the main safety case parts, Objectives or Requirements, Evidence and Arguments. The context of this section is not a simple repetition of the previous two sections but goes into much more technical detail;

5. Related Safety Documents: This section provides references to all other related safety documents and documented processes used in support of the development of the safety case.

For clarity of presentation and ease of appraisal, the System Safety Case (SSC) is structured as follows:

*VOLUME 1:* Presents a statement of safety objectives, confirmation if these have been achieved, and high-level summary of safety arguments and any outstanding safety issues for the railway system configuration stage.
It contains the principal record of the safety arguments for the changes being implemented, and where they have a safety implication on the Project.
The Vol.1 is subject to endorsement, providing a key document input, by the acceptance bodies.

*VOLUME 2:* Contains the safety arguments. It is structured around the core hazards and related safety goals and contains all the key Technical Safety Reports (TSR).

*VOLUME 3:* Contains the 'evidence of safety' in terms of compliance statements, including references to the key Verification and Validation (V&V) records, traceable audit information and close-out reports, Test Results, etc which support the Vol.1 and Vol.2 entries. This volume also refers to all other safety related documents, supply chain safety cases, Audit Reports, Test Logs, etc.

The sections for Vol. 1 of the SSC are as follows:

1. List of References
2. Executive Summary - This section provides a high level SSC status overview:
    a. Purpose and scope of submission for V2.1;
    b. Status of the SSC and related activities including the status of previous submission;
    c. Outline claim of the submission;
    d. Status of identified programme risks related to the SSC delivery;

   e. List of sections altered from previous submission;
3. Introduction - This section consists of following subsections:
   a. Objectives of the System Safety Case;
   b. Aims of this volume;
   c. Scope of the System Safety Case;
   d. Definitions;
   e. Abbreviations;
4. Proposed Submissions - This section presents the proposed submissions of the SSC.
5. System Definition:
   a. Overview of VLU;
   b. System Outline;
6. Structure and Status of the System Safety Case - This section consists of the following subsections:
   a. Position of this Document;
   b. Format of the SSC;
   c. Status of the SSC;
   d. Changes from previous submission;
   e. Comparison with EN50129;
7. Methodology - A summary of the methodology followed during development of the SSC is outlined in this section, as follows:
   a. Safety Goals;
   b. Method of Safety Analysis;
   c. Structure of Safety Argument;
   d. Change Safety Analysis process;
   e. Role of ISA;
8. Safety Justification - This section provides an overview and summarises the status of the safety arguments for the V2.1 project stage, to be detailed in Volume 2.
   a. Completeness of the coverage of the risk;
   b. Train accidents;
   c. Movement accidents:
   d. Non-movement accidents;
   e. Complete set of sub goals;
   f. Safety Arguments Status Report;
   g. Outstanding Safety Issues – Summary;
   h. Conclusions;
   i. Overall GSN;
9. Next submission - The scope and objective of the next SSC submission is outlined in this section;
10. Implementation –Delivery:

   a. Responsibilities;
   b. Liaison with LU;
   c. Liaison with Supply chain;
11. Appendix I: ADCs;

The following sections comprise Vol. 2 of the SSC:

1. Overview:
   a. Scope;
   b. Aim;
   c. Outline of overall safety argument;
   d. Structure of SSC Vol 2.;
2. System Definition:
   a. System description;
   b. Operational envelope of the system;
3. Quality Management report:
   a. Roles and Responsibilities;
   b. Lifecycle Issues;
   c. Standards;
   d. Audits and Assessments;
   e. Supplier management;
   f. Controls;
   g. Configuration Management;
   h. Project Quality Management Training;
   i. Outstanding Issues;
4. Safety Management Report:
   a. Roles and responsibilities;
   b. Safety lifecycle;
   c. Safety analysis;
   d. Safety Standards;
   e. Safety audit and assessment;
   f. Supplier management;
   g. Safety controls;
   h. Configuration management;
   i. Project safety training;
   j. Outstanding issues;
5. Core Hazard/Safety Goal:(NOTE: This section is repeated for each Safety Goal):
   a. Scope;
   b. Description and Process model;
   c. Change Safety Analysis;
   d. Derivation of Safety Acceptance Criteria;

   e. Safety arguments;
   f. Outstanding Safety Issues;
   g. Related Safety Documentation;
   h. Conclusion;
6. Appendix I: Assumptions, Dependences, Caveats;
7. Appendix II: Migration Plan;
8. Appendix III: Project Assurance Plan.

The following sections comprise Vol. 3 of the SSC:
1. Overview;
2. Subsystems: summary of safety documentation/evidence and its status.

The safety case structure recommended in EN50129 is widely recognised as representing good practice in the production of safety cases, and is used well beyond the Standard's nominal scope of electronic systems for railway signalling. Figure 7-14 shows how the structure adopted for the SCC encompasses all the material required by EN50129, and builds into an equivalent structure.

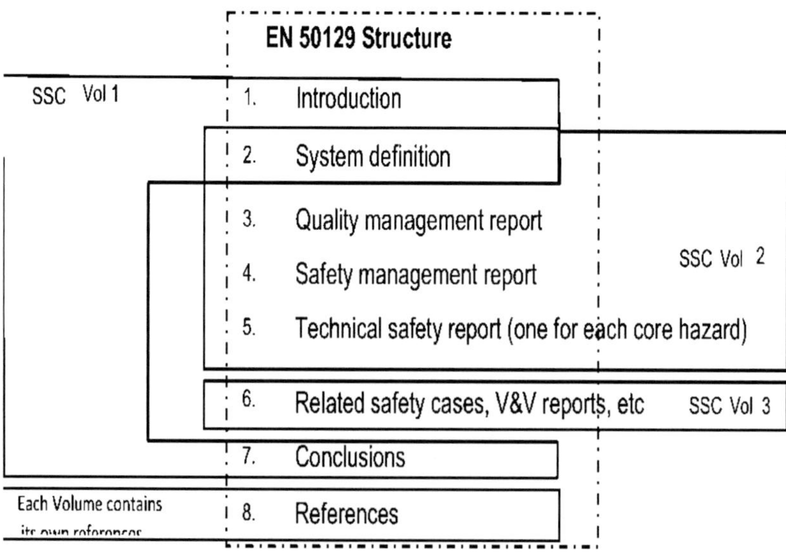


Figure 7-14: SSC mapped onto EN50129

This structure is compatible with the safety case format recommended by CENELEC EN50129 standard, but supports the focus on the control and management of the hazards and consequences.

### *Victoria Line Upgrade Programme – System Safety Case*

The author structured the System Safety Case (SSC) as outlined above. The heart of the SSC is Volume 2 and the outline of the VLUP System Safety Case is presented on Figure 7-15 below.

As the management organisation of the VLUP is relatively simple, there was no need for a GSN presentation of the sections dealing with Quality management and Safety management. The GSN has been developed for each safety goal/core hazard. As an example, a high level GSN for "Collision between trains" core hazard, with the related top safety goal being "Risk of collision between trains meets criteria is presented on Figure 7-16.

The process model is used to set a context for the safety case argument and to focus specific justification for each of the main areas of concern, as indicated by the GSN. For the relevant phase of the project, testing of the new trains in traffic hours, the only elements of the overall process for train movements represented by the process model Figure 7-5, which are changed from the existing Victoria Line, are those associated with the introduction of the new train and the signalling system. The changes comprise:

1. Introduction of new train and signalling system;
2. Modification to existing signalling system to provide interface to new system;
3. Provision of co-acting lineside signals where needed for sighting from the new train;
4. Changes to the driver's tasks associated with the design of the new train and signalling;
5. Changes to activities of signalling, track and rolling stock maintainers.

The outline of the safety argument is as follows:

1. The new train and signalling system will behave correctly providing they are not disturbed and providing they do not receive inputs which could cause unsafe behaviour;
2. The potential disturbing factors in the environment of the Victoria Line will not cause unsafe behaviour of the new train and signalling system;
3. The probability of inputs or actions which would cause unsafe behaviour being applied to the new train or signalling system is

consistent with the safety acceptance criteria;

4. The introduction of the new train and signalling system will not adversely affect the behaviour of the existing trains with regard to the risk of collision.

In addition, special consideration is given to two particular operational activities; moves towards buffer stops and the rescue of failed trains. The risk of train collision is made equal to or lower than the value derived from the LU QRA model by the combination of technology developed to appropriate levels of safety integrity, established good practice in engineering, and the application of effective safety management to operations.

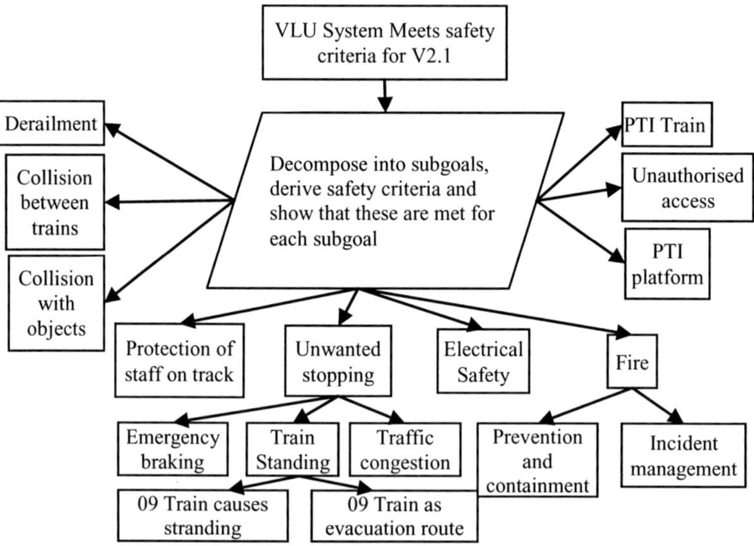


Figure 7-15: VLUP GSN- Top level

Detailed arguments relating the safety justifications with evidence and providing the logical framework have been developed for each of the safety goals from the GSN. The conclusion of the System Safety Case was that the safety risk is not materially impacted by the operation of the pre-series trains in passenger Traffic Hours, and the residual safety risk has been reduced to being as low as reasonably practicable. This was subject to:

1. The imposition of a number of identified restrictions and;
2. Confirmation of the conclusions in the system safety case to be gained upon acceptance of the supply chain safety cases by the programme.

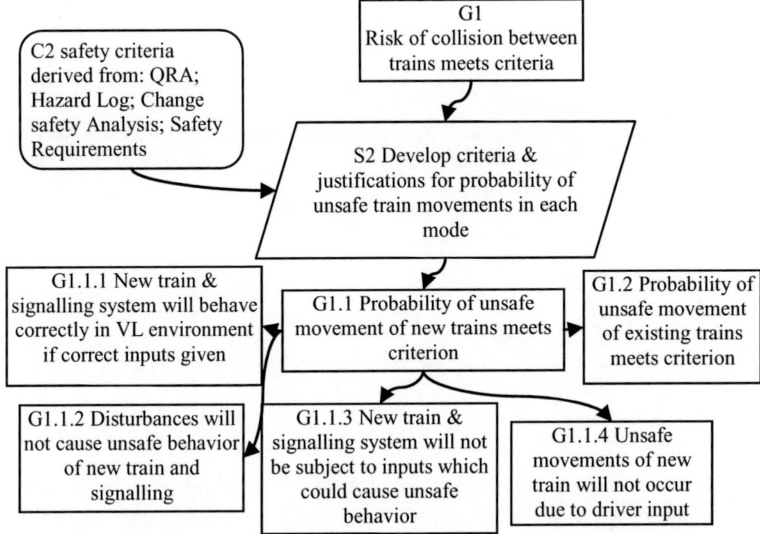


Figure 7-16: Collision between trains GSN (VLUP System Safety Case)

### *Subsurface Lines Upgrade Programme – Hierarchy of Safety Cases*

As already outlined, the Subsurface Lines Upgrade Programme (SUP) delivery organisation is hierarchically structured with many (40 projects) delivering at different timescales to a common objective at a number of migration stages. This arrangement dictates a hierarchical structure and management of the delivery of the safety cases in support of the projects, and the overall programme as illustrated by Figure 7-2, Figure 7-3, and Figure 7-4 earlier. A model of the hierarchy of safety cases, which the author established, is illustrated by Figure 7-24 below.

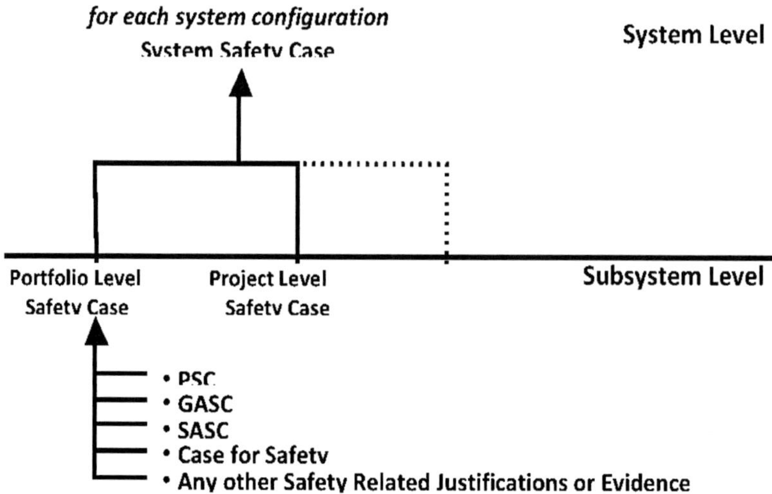


Figure 7-17: Hierarchy of safety cases and justifications/evidences

The subsystem level safety cases are delivered and submitted for approval as and when required prior to individual projects' works being delivered and most often in advance of the system safety case delivery. Some of the project level deliverables are completed at the same time as the system commissioning and are, therefore, delivered and submitted at the same time with the system safety case, but to different approvals/acceptance bodies. However, in order to secure delivery of the complete safety argument, the system safety case structure and fact net must be completed well in advance, with all the subsystem level safety arguments relevant to the system level argument identified and "completed" as part of the subsystem level safety case's delivery and acceptance process.

This is illustrated by Figure 7-18 and Figure 7-19 below (for clarity both figures have been partitioned into a number of figures below). The system level safety argument for Train Movement Accidents, outlined by the GSN (Figure 7-18), covers the complete system. Each of the goals in this particular example is influenced by system changes being delivered by more than one project, as illustrated by Table 20 below.

| Goal | Goal Description | Signalling Immunisation | Legacy Signalling | Signalling Contactor | Train Arrestor Project | Maximum Safe Speed | Rolling Stock | NDUP Project | Structural Analysis Project | ICSS Project | OPO TT CCTV Project | ELLCCR | Stopping Mark Project | DC Traction Power Upgrade | Platform Lengthening |
|---|---|---|---|---|---|---|---|---|---|---|---|---|---|---|---|
| M1 G902 | The risk that the Earthing & Bonding arrangements are incompatible with the existing railway is reduced in line with risk principle | • | • | • | • | | • | • | | • | • | | | • | • |
| M1 G903 | The risk of the movement authority not being visible to the train operator in all rolling stock is reduced in line with risk principle | | • | | • | | • | • | | | | | | | |
| M1 G904 | The risk that the increased length of S Stock is not compatible with signalling layout and clearances is reduced in line with risk principle | | • | | • | | • | • | | | | | | | |
| M1 G910 | The risk that any train is not detected by the Signalling System is reduced in line with risk principle | • | | | | | • | | | | | | | | |
| M1 G916 | The risk that the driver/staff can introduce an unsafe state in the system is reduced in line with risk principle | • | • | | | | • | • | | | | | | | |
| M1 G918 | The risk an unsafe state is introduced in maintainer equipment is reduced in line with risk principle | • | • | | | | • | • | | • | • | | | | |

| Goal | Goal Description | Signalling Immunisation | Legacy Signalling | Signalling Contactor | Train Arrester Project | Maximum Safe Speed | Rolling Stock | NDUP Project | Structural Analysis Project | ICSS Project | OPO TT CCTV Project | ELLCCR | Stopping Mark Project | DC Traction Power Upgrade | Platform Lengthening |
|---|---|---|---|---|---|---|---|---|---|---|---|---|---|---|---|
| M1 G922 | The risk that any objects (e.g. trackside, platform mounted or the platform, structures etc) infringes the Kinematic envelope of the rolling stock is reduced in line with risk principle | • | • | | • | • | • | • | | • | • | | | • | |
| M1 G957 | The risk that any products/ subsystems/systems cannot be supported over the complete lifecycle is reduced in line with risk principle | • | | • | | | | | | • | | • | | • | |
| M1 G958 | The risk that any products/ subsystems/systems cannot be maintained over the complete lifecycle is reduced in line with risk principle - (includes equipment to be located in a position where it can be maintained in safety is reduced in line with risk principle) | • | • | • | • | | • | • | | • | • | • | | • | |

**Table 7-20: Apportionment of responsibility for engineering safety management activities and delivery of evidence across the programme work packages**

The GSN elements coloured in blue are the goals related to the scope of change delivered by the Immunisation portfolio projects.

Following on from that, the Immunisation portfolio GSN and the safety case were produced, demonstrating the justification for safety and providing the evidence in support of the safety argument. This is illustrated by Figure 7-19 below.

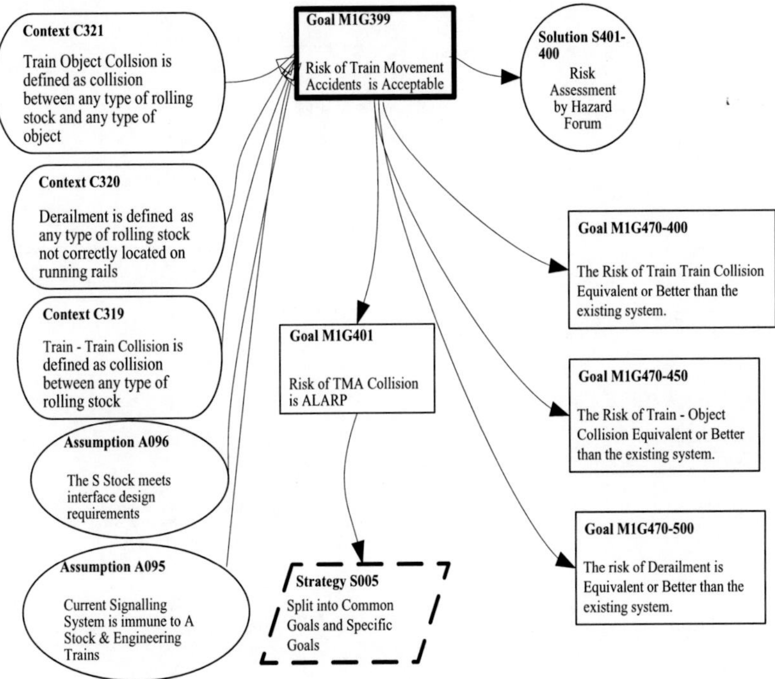


Figure 7-18/1: SSL system level train movements accidents GSN ($1^{st}$ level)

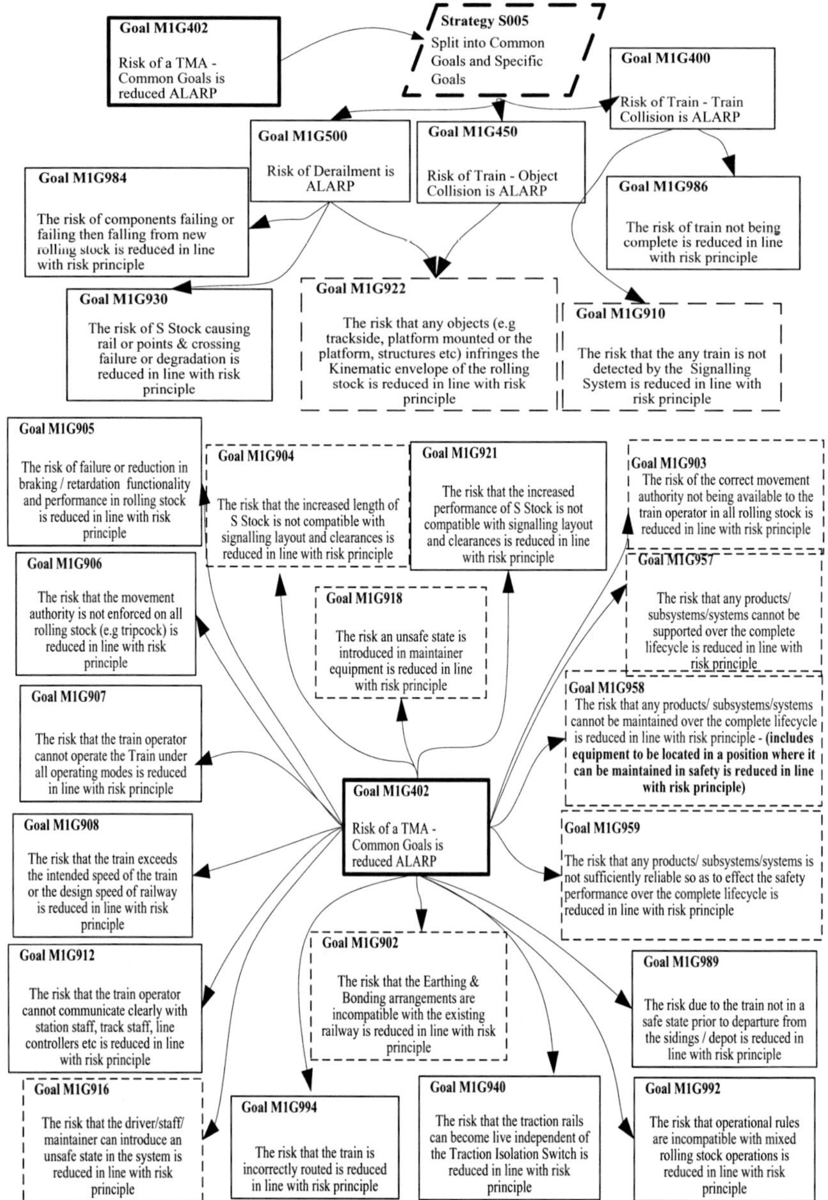


Figure 7-18/2: SSL system level train movements accidents GSN (2nd level)

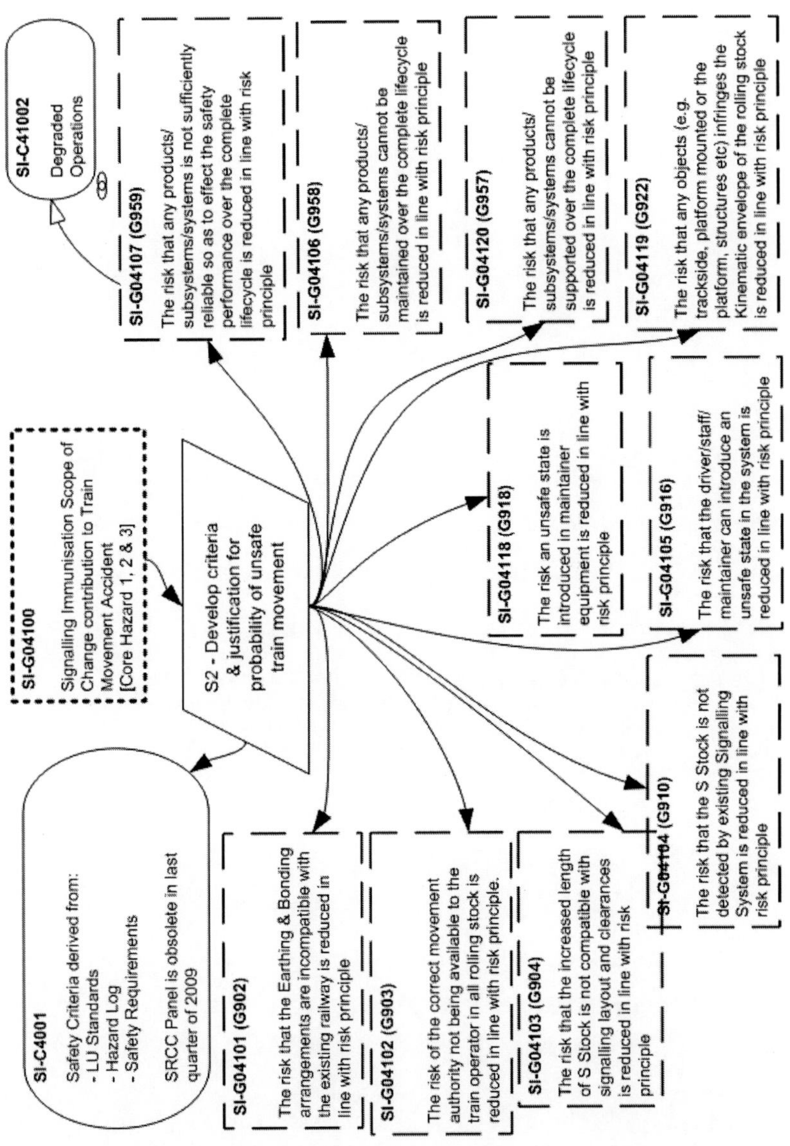


Figure 7-19/1 (previous page): SSL project portfolio (subsystem) level train movements accidents GSN (1st level)

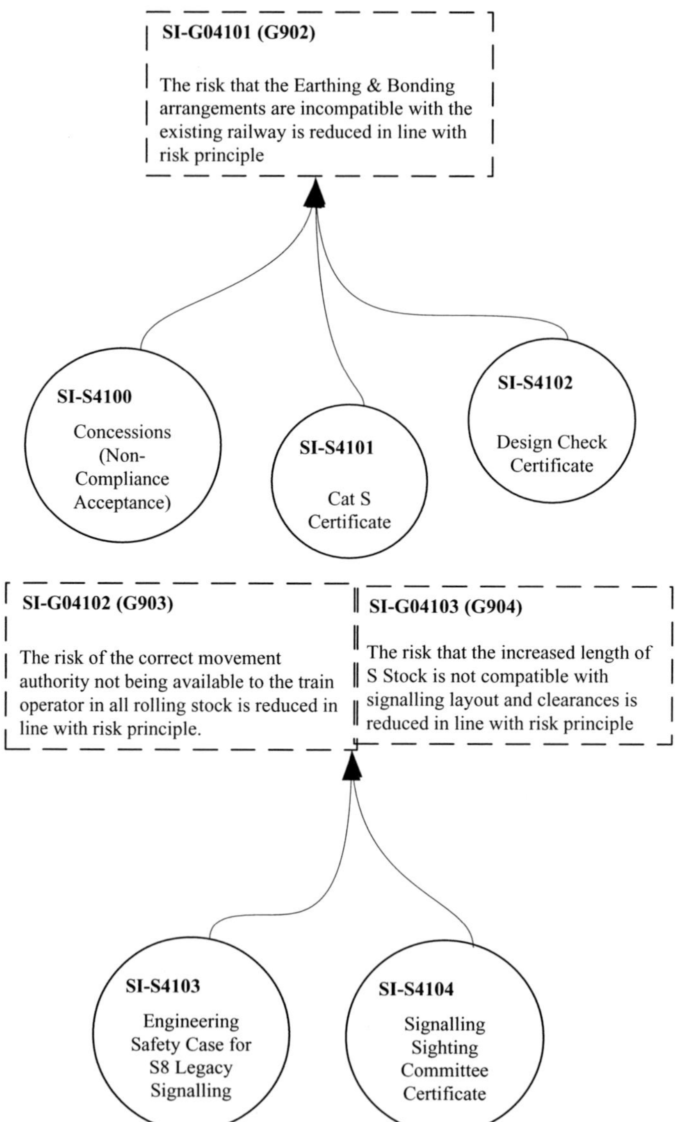


Figure 7-19/2: SSL project portfolio (subsystem) level train movements accidents GSN (2nd level)

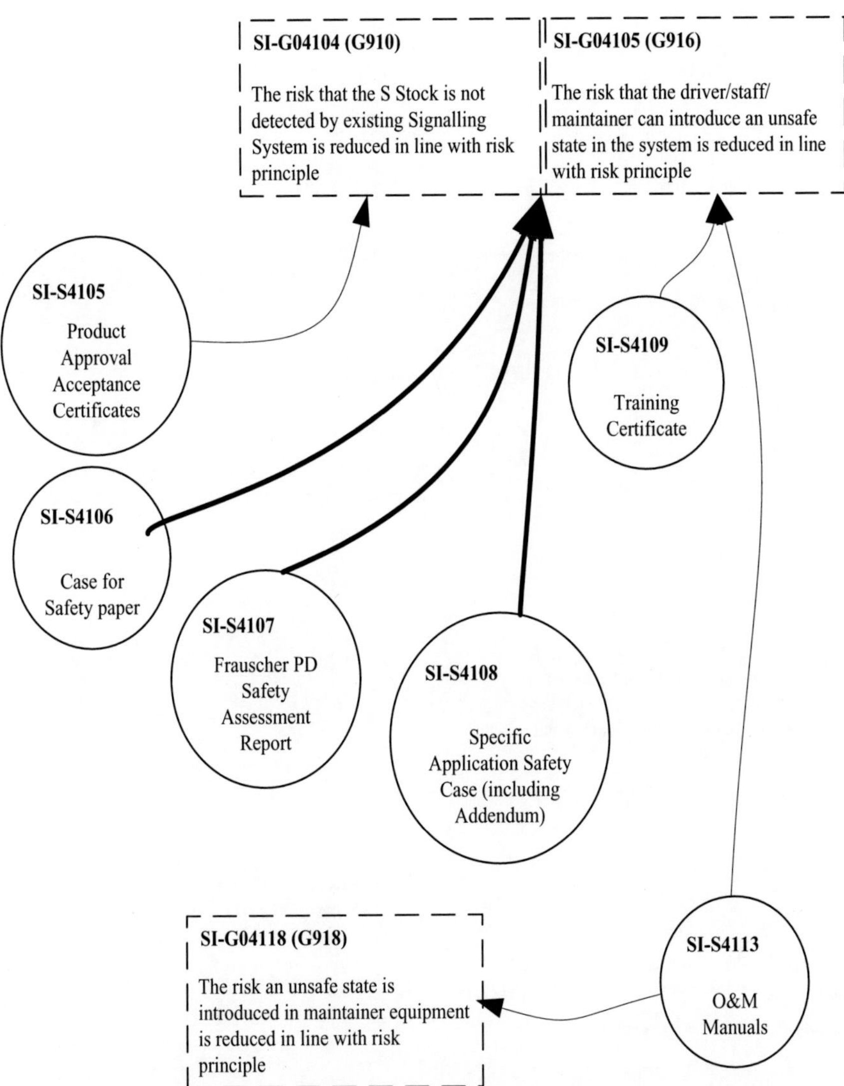


Figure 7-19/3: SSL project portfolio (subsystem) level train movements accidents GSN (2nd level)

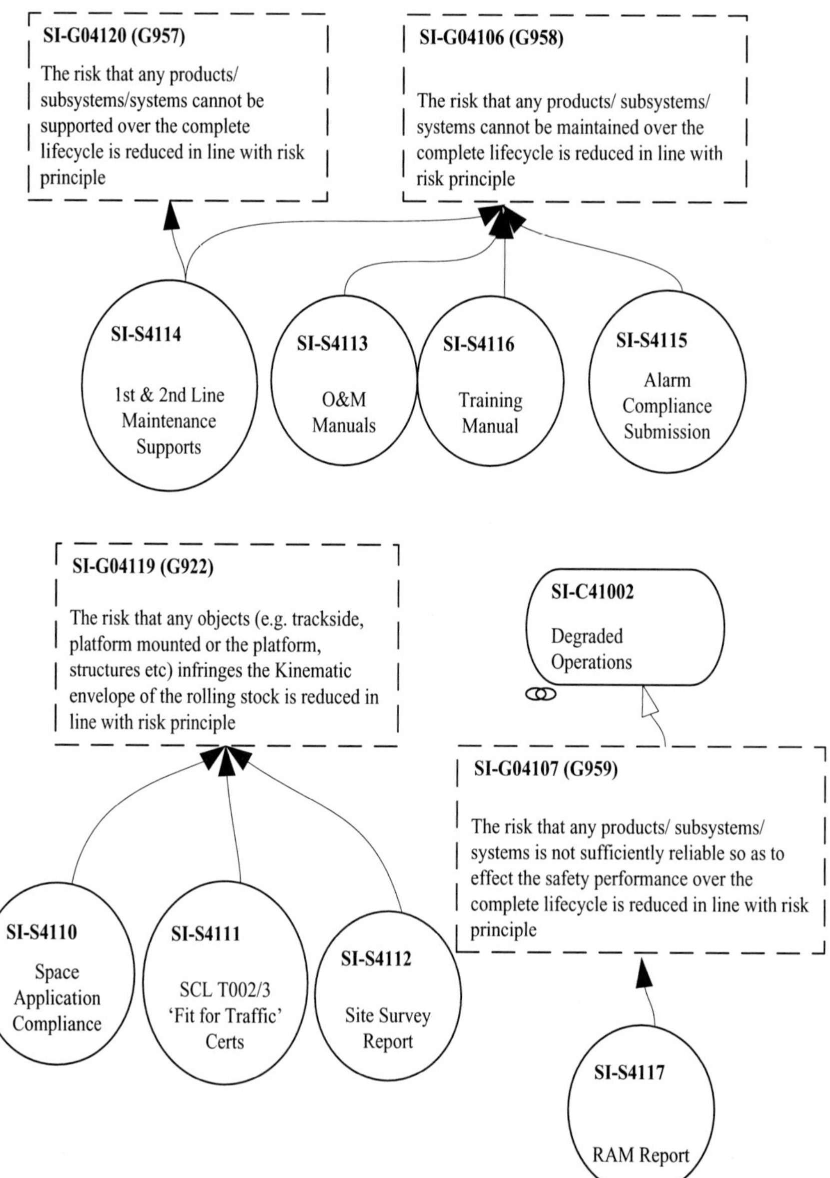


Figure 7-19/4: SSL project portfolio (subsystem) level train movements accidents GSN (2nd level)

The goals (coloured blue) correspond directly to the system level goals, and represent the contribution to the risk from the Immunisation Portfolio implemented changes to the system safety performance. For each of the goals, a remit was produced, agreed with, and passed to the actionee.

This process supports the assurance of completeness in the safety argument and justification. In support of the clarity of delivery and structure of the documentation, the author produced the hierarchical document tree, an example of which for the system level and the Immunisation Portfolio level is shown by Figure 7-20 below. All of the arguments and evidence at project level is summarised in the project safety cases as presented in listed documentation.

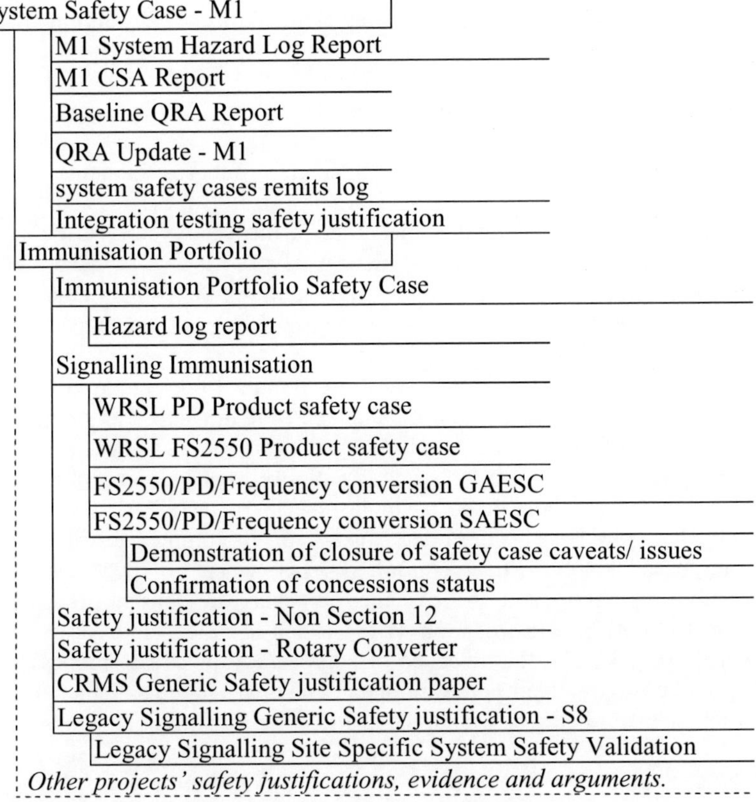


Figure 7-20: SSL System safety argument document tree

## 7.9 Assertion of completeness of analysis and the assessment of uncertainties

An approach to the assurance of the completeness of the system understanding and the completeness and correctness of the Change Safety Analysis has been discussed earlier, and the same processes and methodologies were (and still are) successfully applied on both VLUP and SSL. This will not be further discussed here.

However, during the work on VLUP and SSL an additional type of completeness argument has been identified as required.

### 7.9.1 Completeness of the safety arguments/facts net

As already mentioned earlier, the completeness of the safety arguments net, can be classified into three kinds:

1. Wholeness of lines of reasoning explored;
2. Sufficiency and Completeness of evidence provided in support of these arguments;
3. Certainty of the evidence provided in support of these arguments.

Wholeness of the logical fact net, that is the safety case, is being assured by adherence to the Change Safety Analysis, Risk Assessment and Management process for its development. In addition, the fact net must be transparent throughout its own life cycle, involving all the necessary competences in its development and review.

The use of GSN, or some other form of graphical presentation of the fact net, supports the verification of wholeness of the safety arguments. Substantiation of Sufficiency and Completeness of evidence differs depending on the subject, varying from simple well understood problems, for which the evidence can be a simple check, or a 100% test of all functionalities provided by the relate subsystem, to complex distributed, software based safety critical control systems, where a 100% test is practically impossible and the evidence is likely to consist of some practical testing (with justification of the tested sample being based on statistical analysis) and the evidence of adherence to adequate processes including the use of advanced development and analysis tools. Whichever type of evidence is being provided, the justification for sufficiency must be made.

Certainty of evidence depends on the type of evidence provided. In the case of proof based on a physical measurement (in a very general sense), the certainty is relatively easy to prove through use of well-known methods for the determination of the precision of measurement. However, in the case of non-measurable, intuitive, evidence, typically related to proof of adherence to processes and application of good practice and standards, it is much more difficult to determine the level of confidence in available information. The usual method to support this kind of evidence is by an audit and the monitoring of the initial failure rates through a FRACAS/DRACAS system for more frequent than expected failures of the system as an indication of inadequate processes in place or existing mitigations, although being implemented, not being effective.

## 7.10 Management and Reporting

### 7.10.1 Management

The planning and management of safety management activities is no different to that required for the management of any other engineering undertaking, apart from that the following activities are specific to engineering safety management (common planning and management will not be further discussed):

1. Management of the hazard logs;
2. Management of the operational restrictions;
3. Management of the safety case delivery.

These are related and, to a great extent, overlap. The hazard log, as already mentioned, holds a record of all the hazards and related change safety activities. Apart from making sure that all the records are kept up to date and configuration controlled, which will be discussed later, the hazard log is a tool for the control and management of hazard log actions. Each action needs to be allocated to an owner with a clear and concise description of the action and what the action is expected to deliver.

Operational restrictions need to be managed and controlled with the same rigour as the hazard log. During the identification and analysis of the restrictions, it is necessary to identify the evidence required and the "authority" to remove or relax the restriction. No restriction should be removed without the documented agreement from the identified authority.

Most often, delivery of the safety case is dependent on input from people or teams other than the safety case author or the engineering safety team. On SSL and VLUP programmes, delivery of the safety cases has been managed through the remits. At project level, Project Managers are responsible for the delivery of all documentation on projects including that required to support delivery of the system safety case.

It is the Project Manager and the delegated Project Engineer's responsibility to ensure that evidence is collated and controlled that will support all of the arguments required for the safety case, including the maintenance and operations readiness and RAM performance acceptability.

The system safety team identifies the safety arguments and evidence requirements and develops remits for delivery by the Project Manager or Engineer in support of the system safety case, see Figure 7-21.

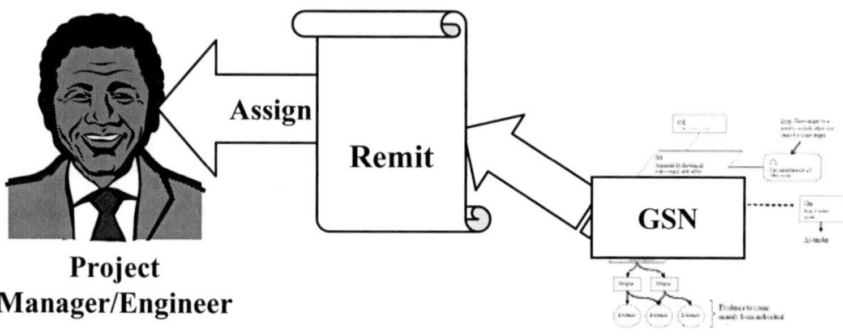


Figure 7-21: Delivery management of the Safety Case

The management of a safety case via remits, allows for the reporting against the remit's delivery, as KPIs and monitoring of progress.

In order to de-risk the assurance regime, for each system configuration, several versions of the system safety case were submitted for review well in advance of the commissioning (this strategy is still being followed on both projects). As represented by the Figure 7-22 below, the following approach was successfully implemented:

1. As early as possible, a template document, indicating structure and main areas of discussion, is submitted for agreement and approval;
2. A first draft with 85% to 95% of the requirements identified, all evidence items intended to support these identified, and related arguments completed subjected to review and approval by all stakeholders prior to final commission enabling submission;
3. A system safety case with all the requirements identified, all the evidence items intended to support these identified and the most available, related arguments completed submitted for approval to allow commissioning to happen;
4. Due to the complexity of the changes introduced to the system, a post commissioning system safety case update is submitted in order

to close out any outstanding issues and enable the review of the adequacy of implemented mitigations and temporary mitigations.

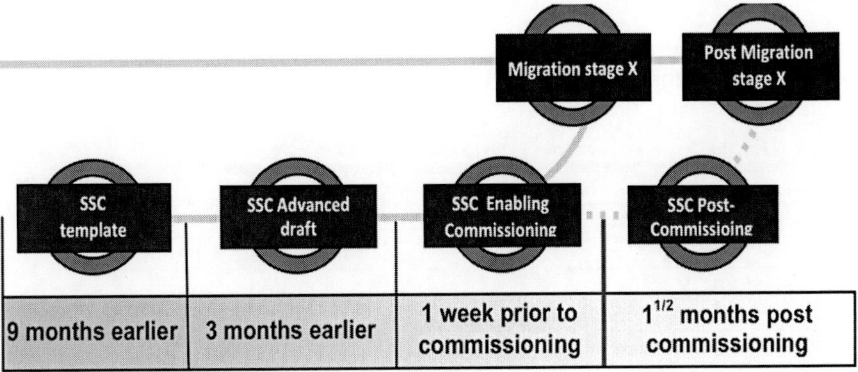


Figure 7-22: Typical programme of submissions.

### 7.10.2 Reporting

The author identified five types of reporting during the implementation of the framework:

1. Analysis reporting. These reports present the analyses carried out and are usually submitted for approval and acceptance in preparation for the commissioning of the system change. These reports should have the same general structure as the safety case; objective or requirements, arguments, evidence and context;
2. Analysis records are used to create an auditable trial of evidence gathering in support of the analysis reports. These are used as evidence, referred to and from the safety case;
3. Management-planning. Necessary safety management activities should be identified early in the project, planned and agreed within the project as well as with the regulatory and acceptance bodies. The plans need to be recorded and regularly updated to support an auditable trail and continuity throughout the life of the project;
4. Management-implementation. Implementation of actions resulting from the Change Safety Analysis and management activities, should be controlled through remits agreed between the actionee and the safety team;
5. Management-progress monitoring. On any project it is necessary to monitor progress of delivery. With regards to the change safety

management process, progress monitoring should be done through monitoring of the remits.

List of reports that have been identified and are being produced is presented in Table 7-21 below. Directly related reports are indicated by outline numbering.

| *Report* | *Report category* |
|---|---|
| 1. Safety Case: | Analysis reporting |
| 1.1 Safety Cases Remits; | Management-implementation |
| 1.2 Remits Progress Reports; | Management-progress monitoring |
| 2. Other safety cases, justifications & documents; | Analysis reporting |
| 3. Hazard Log Report: | Analysis reporting |
| 3.1 Action Remits; | Management- implementation |
| 3.2 Hazard Forum Minutes; | Analysis record |
| 3.3 Safety Requirements Forum Minutes; | Analysis record |
| 3.4 Actions progress report; | Management-progress monitoring |
| 4. Restrictions Register; | Management- implementation |
| 5. Hazard report; | Management- reporting |
| 6. CSA Report; | Analysis reporting |
| 6.1 CSA Briefing; | Analysis record |
| 6.2 ICSA Records; | Management- implementation |
| 6.2.1 Change Requests; | Analysis record |
| 6.2.2 ICSA Briefing; | Analysis record |
| 7. QRA Report: | Analysis reporting |
| 7.1 QRA Models; | Analysis record |
| 7.2 QRA Data log; | Analysis record |
| 7.3 QRA Change Register; | Analysis record |
| 8. Process Models Report; | Analysis reporting |
| 9. QRA Development Plan; | Management-planning |
| 10. Safety Audit Plan; | Management-planning |
| 11. Acceptance & Certification Strategy; | Management-planning |
| 12. Engineering Safety Management Plan. | Management-planning |

**Table 7-21: Reporting - Change Safety Management process**

## 7.11 Configuration Control, Continual Appraisal and Knowledge Management

Within the change safety management process a number of different configurable data items exist:

1. System change data:
    a. System Change scope and scope changes;
    b. System configuration data;
    c. Assumptions and Dependences;
2. Safety analysis data:
    a. Hazard log data;
    b. Risk assessment information (qualitative and quantitative);
    c. QRA data;
    d. Incidents and Accidents data;
    e. Safety related documentation;
    f. Operational Restrictions/Caveats;
3. Safety performance monitoring data:
    a. FRACAS/DRACAS systems;
    b. Incidents reporting system;
4. Management data:
    a. Planning;
    b. Implementation and progress monitoring.

Relationships between these are described in Figure 7-23 below. System change data enables and provides input into the system analysis. All of these must be strictly configuration controlled, because the change to the system data may have an impact upon the system safety performance. Frequent appraisal of how correct and current the information is , is crucial. Safety analysis and safety performance monitoring data are critical to the development of a suitable, applicable and convincing safety case. These must be configuration control as well, but in addition to that, this data should be subjected to continual appraisal, throughout the life of the project. Once the project is completed, it shall be handed over to the infrastructure owner, it needs to be kept live and up to date until the system is decommissioned or upgraded. The management data is used to record the processes and monitor the progress towards a successful delivery of the safety argument. Therefore, this data does not have to be strictly configuration controlled, but needs to be sufficiently current to enable appropriate progress monitoring and appraisal as required.

In line with the recommendation from the Yellow book (RSSB, 2007) some of this information is contained within the hazard log (as described earlier) and configuration control of it is managed through the parts of the hazard log itself.

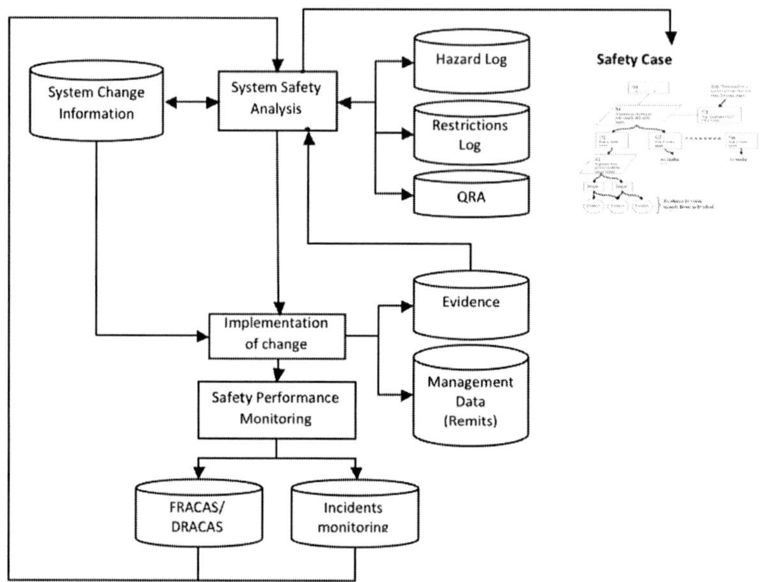


Figure 7-23: Configurable data items

*Journal* The Journal describes all amendments to the Hazard Log, in order to provide a historical record of its compilation and provide traceability.

For each amendment it contains:

1 The date of the amendment (not necessary if diary format is used);
2 A unique entry number;
3 The person making the amendment;
4 A description of the amendment and the rationale for it;
5 The sections in the Hazard Log that were changed.

*Directory of safety related documents.*_The Directory, provides an up-to-date reference to every safety document produced and used by the Project.

The documents referred to include the following, where they exist:

1 Safety Plan;
2 Safety Requirements Specification;

3 Safety standards;
4 Safety Documents;
5 Incident/accident reports;
6 Analyses, assessment and audit reports;
7 Safety Case;
8 Correspondence with the relevant Safety Approvers.
9 For each document the Directory includes the following:
  a. A unique reference;
  b. The document title;
  c. The current version number and issue date; and
  d. The physical location of the master.

*Hazard Data (including safety requirements).* Every identified hazard is recorded. For each hazard, the information listed below is recorded as soon as it becomes available and is strictly configuration controlled.

Data collected during Hazard Analysis and Risk Assessment is contained within the Hazard Log:

1 A unique reference;
2 A brief description of the hazard including the system functions or constituents affected and their states, as they represent the hazard;
3 The causes identified for the hazard;
4 A reference to the full description and analysis of the hazard if not contained in the hazard log;
5 Assumptions on which the analysis is based and limitations of the analysis;
6 The severity for the related accident, the likelihood of the hazard occurring and the likelihood of an accident occurring with the hazard as a contributing factor;
7 The predicted risk associated with the hazard;
8 Target likelihood for its occurrence;
9 A reference to all Safety Requirements associated with this hazard or with related causes or protection measures;
10 A reference (and description) to the system interface or object:
  a. Whose change is the source of cause or hazard;
  b. To which a protection measure is related;
11 A reference (and description) to related QRA elements (base and top events, gates, etc);
12 The status of the hazard;
13 If the hazard is not closed or cancelled, then the name of a person who is responsible for progressing it towards closure;

14 A description of, or a reference to, the action to be taken to remove the hazard or reduce the risk from the system to an acceptable level. This includes:
   a. A statement as to whether the hazard has been avoided or if it requires further action (with a justification if no further action is to be taken);
   b. Details of the risk reduction action to be taken;
   c. A discussion of the alternative means of risk reduction and the justification for actions considered but not taken;
   d. A comment on the need for accident sequence re-evaluation following risk reduction actions;
   e. A reference to any design documentation that would change as a result of the action;

15 Details of any residual risk or reference to another entry in the hazard log that contains details of the residual hazard.

*Incident Data.* At the time of writing this book, no incidents occurred. However, in future, all incidents that have occurred during the life of the system or equipment are to be recorded here, identifying the sequence of events linking each accident and the hazards that caused it.

For each incident the following will be provided:

1 A unique reference;
2 A brief description of the incident;
3 A reference to a report describing an investigation of the incident;
4 A description of any action taken to prevent recurrence, or justification of the decision not to take any.

*Consequence/Accident Data.* Every identified possible consequence is recorded; including the possible sequences of events linking identified consequence with the hazards that may cause it.

For each consequence the following is provided:

1 A unique reference;
2 A brief description of the potential consequence;
3 A reference to a report giving a full description and analysis of the consequence sequence;
4 A risk assessment information (either comparable, using earlier outlined methodology or quantified data); and
5 A list of the hazards and the associated consequence sequences that could cause the consequence.

*Operational Restrictions* are being managed by a centralised Restrictions register. As operational restrictions are in fact temporary mitigations these need to be treated with the same rigour as other hazards. Table 7-22 is an example of the record.

<table>
<tr><td>Ref</td><td>Description</td><td>Group</td><td>Rationale</td></tr>
<tr><td>R014</td><td>The full performance key switch shall only be operated (to select full performance) whilst in an Engineer's Current Area. At all other times the key switch shall be OFF.</td><td>Rollin stock - power</td><td>Full execution of test</td></tr>
<tr><td>Action</td><td>Actionee</td><td>Removal Criteria</td><td>Removal Authority</td></tr>
<tr><td rowspan="5">Ensure train is in Inter-Running mode when outside Engineer's Current Area.</td><td>Lead Train Master.</td><td rowspan="5">Conclusion of all tests requiring operation at "Full Performance" and removal of key switch mode.</td><td>Head of Rolling Stock Engineering SUP, Transportation</td></tr>
<tr><td>Status</td><td>Evidence or Action</td></tr>
<tr><td>Open</td><td>#N/A</td></tr>
<tr><td colspan="4">Notes</td></tr>
<tr><td colspan="4">Revised 16/11/2009 following installation of Mod No. SSL1509.</td></tr>
</table>

**Table 7-22: An example of the restriction's register record**

*QRA* related information can be split into two categories; data and logic. Model data needs to be regularly reviewed and confirmed against the latest incident data. The logic of the QRA model, once the model has been completed (including review for correctness), needs to be reviewed after each change to the system. Usually, tools used for QRA modelling have data storage facilities that can be used to save different versions of the model. On VLUP and SUP, Fault Tree + software has been used for modelling. All of the modelling data is stored within the tool's data repository. The configuration control is implemented through the tool following a change control process.

*FRACAS/DRACAS*: Failure/Defect Reporting and Corrective Action System is in place on both VLUP and SUP. This is a standard process and will not be further disused here, apart from mentioning that it is essential for the information from the FRACAS/DRACAS system to be reviewed against the records in the hazard log and the restriction log to confirm that the mitigations in place are working and are sufficient, and that there are no missing causes in the hazard log.

*System Change Information*: as the system is being changed it is necessary to keep the information about the change up to date and ensure that the system description is complete and correct. The process was already discussed earlier in this book, has been successfully implemented on both projects and is working.

*Incidents monitoring* is implemented across the LU. All of the incident reports are subjected to analysis by the engineering safety team. Any relevant information is assessed, and, if appropriate, used to update the QRA and/or hazard log.

*Remits* are used in support of the management process in order to convey the details of actions that need to be taken by the actionee, including the timescales for delivery. As already explained, the status of remits is used as an indication of progress of the safety case delivery. An example of a remit is presented below in Table 7-23.

<table>
<tr><th>Task Id.</th><th>Context</th><th>Task Description</th></tr>
<tr><td>BTLUP1</td><td rowspan="5">Numerous items in the Hazard Log relevant to train signalling system, all are controlled or mitigated by the measures on which the safety arguments are based.</td><td rowspan="5">Provide regular update on the status of the BTLUP Hazard log including:<br>• clear and unambiguous statement of arguments in support of closure of hazards;<br>• an assessment of residual risk<br>• reference to relate the hazard log entry to process model interface.</td></tr>
<tr><td>SSC Section</td></tr>
<tr><td>Collision and Derailment</td></tr>
<tr><td>SSC reference</td></tr>
<tr><td>New section on hazard logs</td></tr>
</table>

<table>
<tr><th>Delivery</th><th>Milestone</th><th>Owner</th><th>Type of response agreed</th></tr>
<tr><td rowspan="3">dd/mm/yy</td><td rowspan="3">Hazard log report for V2.1 system configuration with all relevant hazards closed</td><td>Name</td><td rowspan="3">LUP: An update to the hazard log report with all hazards closed/conditionally closed.<br>LUP: Confirma that all hazards are transferred from MR to BT.</td></tr>
<tr><td>Status</td></tr>
<tr><td>Draft response received</td></tr>
</table>

**Table 7-23: An example of the remit**

## 7.12 Chapter Conclusions

In this chapter, the author outlined and analysed the results of real life applications of the proposed framework. As part of this work, the author identified some minor gaps in the framework, developed missing processes and methodologies to fill these gaps. These have been tested and analysed as well. Detailed conclusions from this work are presented later.

*You are note defined by your past. You are prepared by your past.*

*—Anonymous*

*During one of many German air raids on Moscow in WW II, a distinguished Soviet professor of statistics showed up in his local air-raid shelter. He had never appeared there before. "There are seven million people in Moscow," he used to say. "Why should I expect them to hit me?" His friends were astonished to see him and asked what had happened to change his mind. "Look," he explained, "there are seven million people in Moscow and one elephant. Last night they got the elephant."*

*—Anonymous*

# CHAPTER EIGHT

# CONCLUSIONS, CONTRIBUTION AND SUGGESTIONS FOR FUTURE WORK

## 8.1 Conclusions and contribution

This section summarises the findings of the research and identifies how this has contributed to the overall knowledge and practice of System Safety. The structure of this section is outlined by figure below.

### 8.1.1 Overall summary

The author completed a thorough exploration of the existing methodologies for safety risk analysis, assessment and processes for the management of engineering safety in the railway industry. The author considered and commented upon a substantial number of methodologies, tools and modelling techniques. The findings of this initial research are outlined in chapters two and three of the book. Following from there, the author used as vehicles a number of projects for further analysis and development of the initial ideas as outlined in chapter 4.

The author identified a number of gaps, in particular in relation to the establishment of a systems based framework for analysis, assessment and management of engineering safety. These are discussed in chapter 5. As part of this work, the author identified and discussed the requirements for a new system based framework. The requirements framework has been structured to correspond to the number of stages within the safety risk analysis and management process.

The framework supports a system based approach to safety analysis and management, and provides complete step by step guidance for:

1. Conceptualisation of the system to support safety analysis;
2. Identification, analysis, assessment and management of safety risks;
3. Construction and management of safety cases.

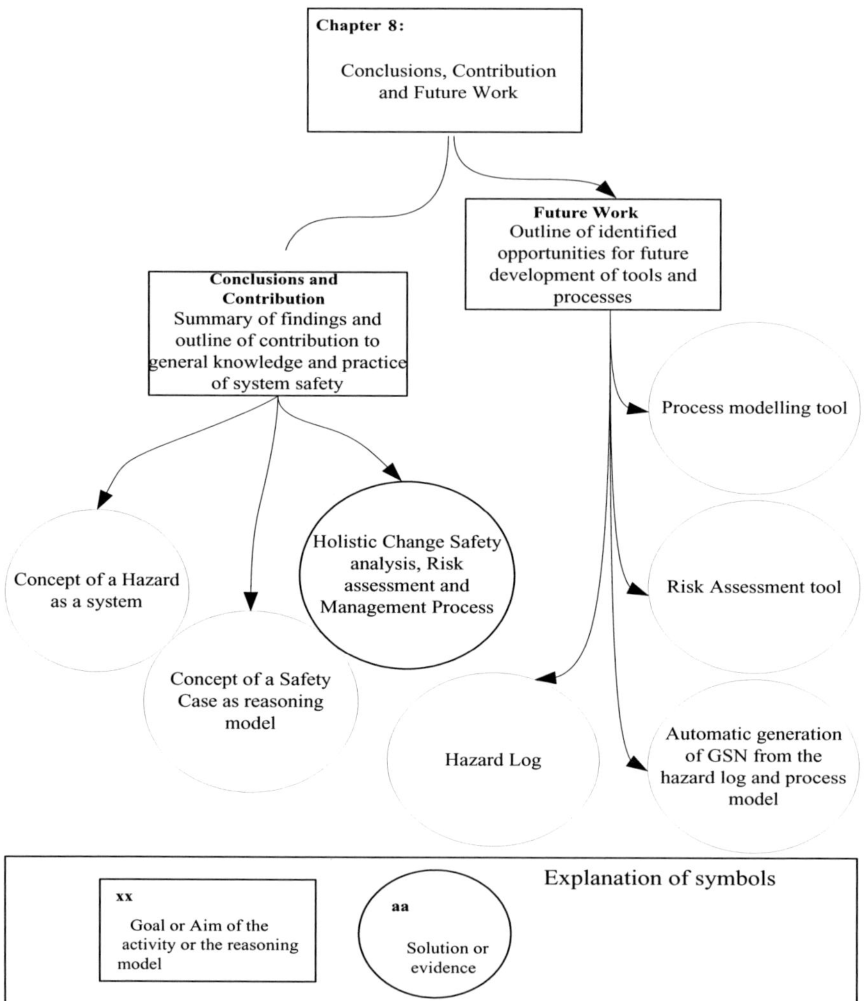


Figure 8-1: Structure of Chapter 8

The author successfully implemented and trialled this framework on two real life railway projects. Each of these stages is considered in turn, and the findings presented here, with a clear indication of the fulfilment of the identified requirement and the contribution to the wider knowledge and practice of engineering safety.

Additionally, the author analysed and further developed the theoretical background in relation to the concepts of "Hazard" and the "Safety case".

### 8.1.2 The concepts of "Hazard" and the "Safety case"

The author analysed the hazard as a system concept and developed existing notions further, identified and defined the attributes, as well as the hierarchical structure of a Hazard as a system.

The author analysed the concept of the safety case as the inquiry model identifying a number of constraints and requirements that a safety case must satisfy. Subsequent to analysis of these, the author proposed a general high level approach to the structure of the safety case as a fact net reasoning model.

This knowledge and understanding informed the later development of the engineering safety analysis and management process and methodologies.

### 8.1.3 Holistic Change Safety Analysis, Risk Assessment and Management Process

The author analysed the existing system safety and system engineering processes and guidance, and proposed a new framework. Firstly the author outlined the vision of safety engineering as an integrated part of the system engineering process, and the assurance of delivering the needed functioning of the four key emerging properties of any system: safety, RAM, Operability and Performance.

Following from that, the author developed and outlined the integrated process for the analysis and management of engineering safety. The process consist of 7 steps, most of which are not newly identified as needed but have not been formalised nor used together within an integrated framework.

As part of this research the author identified requirements for the integrated process, analysed existing methods and processes, identified gaps and outlined the need for a framework in chapter 4 of this book.

***Step 1: System Conceptualisation, Representation and Scoping (System Analysis)***

A number of methodologies for system conceptualisation have been analysed and some of these tested (chapters 2, 3 and 4). The author adopted the methodology initially developed by Short, “Process Models”, formalised it further and integrated it with the rest of the framework.

### *Step 2: Information processing*

The processing of information can be structured in a number of bite sized, manageable activities, to reduce complexity and increase clarity. Following on from this is a summary of the research findings and a contribution statement against each one of these activities.

*Core hazard grouping, Hazard identification and Analysis of causalities and consequences*

A two stage process, Initial Change Safety analysis and Change Safety Analysis, has been developed and implemented. As depicted in Chapters 6 and 7.

### *Risk assessment*

The author undertook a review and analysis of existing risk assessment techniques. The author developed a technique and methodology for the integration of standalone QRA models into an integrated system risk model as well as the specification for software tools in support of this methodology. This work is outlined in Chapter 4, in the sections dedicated to the 1st and 2nd Project.

The author developed a process that brings together the qualitative and quantitative risk assessment methodologies into a framework supporting the production of the safety case and the ALARP argument. This has been applied on the real life project (LU Upgrades) as depicted in Chapter 7. The quantified approach was used to set the safety targets, and to monitor safety performance of the railway throughout the migration whilst in support of the ALARP argument a comparative, qualitative, method was used.

### *Estimation and apportionment of the safety risk to the source causes and failures of preventative measures and risk profiling*

As part of the project depicted in Chapter 4, section 1st Project, the author developed algorithms for the calculation of the contribution of equipment and procedural failures, and human errors, to the total safety risk. This methodology was successfully applied on the project and used to establish the safety targets at the subsystem level for the new train control system. The same methodology was used to develop a multidimensional risk profile.

As part of the LU Upgrade project, the existing LU models have been utilised to define the system level hazard rates and calculate the contribution of the signalling system and the rolling stock related failures to the overall risk.

***Identification of different solutions and mitigation measures (options and impact analysis) and optimisation of the system and selection of most gainful options and mitigations***

The outline of the need to undertake options and impact analysis is defined by industry standards, and the legal framework in the UK. The author developed this framework further, combing the use of quantitative and qualitative analysis and assessment.

The numerical safety targets are set using quantified risk modelling. The impact of the achieved performance of the subsystems on the overall system safety performance is tested using the QRA.

The system level ALARP argument is supported by the ALARP review process developed by the author using the existing comparative assessment criteria. As part of this assessment, if there is a need for any further reduction of risk, different options are assessed and evaluated, taking into consideration all relevant parameters as defined by the framework.

*Assertion of Completeness of System Analysis and Assessment of Uncertainties*

The author analysed the problem of the completeness of the system analysis and with the uncertainty of information and data. The author identified four distinct areas where assurance for completeness and analysis of uncertainty is required.

***Step 3: Derivation and management of requirements***

Existing well-defined process for the derivation of safety requirements have been followed and applied within the new framework. The author analysed the concept of the safety requirement from the system point of view and proposed a minor clarification of the definition and the topology of a safety requirement as a system safety "object" in relation to defence/protection measures.

***Step 4: Presentation of results and construction of safety knowledge net***

The author analysed the reporting requirements in support of the safety related activities and has identified 5 different types of reports.

Of particular importance is the Safety Case. The complexity of modern systems inevitably brings about an increase in the intricacy of the safety case arguments or fact nets. Thus, it is necessary to endeavour to simplify the logic of the safety case fact net for a number of reasons:

1. To provide the focus thus assisting the experts involved in production of the safety case;
2. To assist the understanding of safety within the system thus aiding the reviewers;
3. To ease the management of updates and configuration control of the data, and subsequently changes to the safety case.

A method of achieving these requirements, discussed earlier, has been implemented on both VLUP and SSL.

***Step 5: Management support***

In addition to the usual project management related activities, it has been thoroughly established that the management of the hazard logs, operational restrictions and the safety case delivery, is a vital part of engineering safety management activities.

Regarding a hazard log as a tool for Change Safety Management, Configuration Control and Reporting, the author analysed the problem domain from the system perspective, identified requirements and defined the data model (hierarchy and topology) for the hazard log.

With regards to the management of operational restrictions, the author analysed the problem domain, identified the process requirements, and defined the template for the restrictions register.

Finally, with regards to the management of the safety case delivery, as part of the overall framework the author identified the requirements and defined the processes supporting this activity, including an outline template of a remit document.

For all reports associated with the above activities, the author identified the requirements and defined the template and the context, including a means of measuring progress.

***Step 6: Configuration control***

As an integral part of the research, the author identified the configurable data items, defined the hierarchy and topology of these, and defined the requirements for the configuration management of this data. In completing this work the author used guidance from the industry standards, which provide a high level outline of basic needs for configuration management but do not provide the detail.

***Step 7: Continual appraisal and knowledge update***

Apart from implementing existing guidance as part of the system safety management process framework, the author identified that further research was not required at this time to support the other areas assessed in this book.

## 8.2 Future Work

The system safety analysis and management framework that the author developed has been implemented, and has proved to be both a vital and cost effective process for the assurance of safety.

Dissemination of the good practice within LU continues with the "ESAC College" being used as a vehicle to educate and normalise the effort of both project engineers and managers.

As part of the continuous development and improvement policy the author embedded within the team, we are challenging the framework process and the methodologies as well as deliverables. The change control process is established, and in case a weakness in the process or omission within any of the deliverables is identified, immediate action is taken to identify and implement corrective action.

Most of the methodologies used within the framework are supported by tools; however a tool that could support and automate the integration of the methodologies would be extremely beneficial. At the moment this facility is embedded within the hazard log database, but the process of cross-referencing the hazard constituents to originating process models interfaces, QRA elements, safety requirements repository, GSN elements, related documents, etc is manual, time consuming, and prone to human error.

Some of these facilities are provided in the Integrated Safety Assurance Network (link between the QRA and Hazard log and documents register), and so the logical step forward would be to utilise and extend the capabilities of the ISAE tool.

The author is working on the development of a specification, and subsequently, a business case for the development of a tool (or a toolset) that will provide the following capabilities within the same environment:

1. Process modelling tool:
    a. GUI with symbols tool set;
    b. Object and interfaces database (object and interfaces created and information populated via GUI);
    c. Facility to call up a hazard log user interface from an object or interface symbol, automatically passing cross referencing information to newly created hazard log record and establishing a link between the two data sets;
    d. Reporting, both hard and soft;
2. Hazard log with functionalities described earlier has already been developed. However following functionalities should be added:
    a. Real time aligning and parallel working with the Process Modelling environment;
    b. Automatic generation of the "hazard universe" graphical metta model to support analysis of its structure, constituents and topology;
    c. Automated linkage to requirements management tools (DOORS for example) enabling automatic cross referencing and tracking of progress;
3. Quantitative and Qualitative Risk Assessment. A tool that provides most of the functionalities for qualitative and quantitative risk assessment with linkage to the hazard log already exists. The ISAE tool delivers part of the functionality required to support the framework. Further enhancements are necessary as follows:
    a. Automatic linkage between the "hazard universe" constituents and Quantitative and Qualitative Risk Assessment data constituents;
4. GSN generation tool. The author believes that it is possible to automate the generation of a draft GSN using the structured information in the hazard log. This GSN could only ever be used as a draft for detailed review, but creation of a real-time link between the "hazard universe" constituents and the safety case arguments-evidence net would be extremely valuable to support reviews and sense checks as well as enabling automatic synchronisation between the hazard log (and through hazard log

the process models, QRA, Requirements, Restrictions, actions management and the safety case.

A railway system is a "System of Systems" (SoS) (Karcanias and Hessami, 2008) and (Karcanias and Hessami, 2008). As the theory of SoS advances, the process outlined in this book need to be kept aligned with the latest developments.

A particular limitation of the process is related to lack of temporal properties of the process models. Further development of the process models methodology to cater better for the temporal properties of the system, improving the identification and analysis of emerging properties of complex systems, is a challenge that would be a worthy undertaking.

Initial results from a number of case studies in different industries indicate the benefits that could be achieved from further research and refinement into the area of justifiable spend, and an extension of the research into J-value, carried out by Thomas and Stupples (Thomas and Stupples, 2006), to incorporate societal perception of risk acceptance.

## 8.3 Final Thoughts

Systems, and thus the hazard universe, are forever changing and getting more complex. This is an inevitable part of the world we are in. The work described in this book has smoothed that transition, and helped to make systems and hence our daily lives a little safer.

# APPENDIX A

# ALGORITHMS FOR APPORTIONMENT AND IMPORTANCE IN CAUSE-CONSEQUENCE MODELS

## Apportionment of a virtual consequence to lower level virtual consequences

The apportionment of virtual consequence to the lower level virtual consequence is calculated as a proportion of the Risk contributed to the virtual consequence by the lower layer virtual consequence. Therefore, if $R_{vj}^{l}$ is a risk of the virtual consequence 'j' ($1 \leq j \leq n$) at level 'l' ($1 \leq l \leq n$), then,

$$R_{vj}^{l} = \sum_{i=1}^{i=n} R_{vi}^{l-1} \qquad \text{Equation 5}$$

Hence, the fractional contribution 'A' of the virtual consequence 'x' ($1 \leq x \leq n$) at lower level 'l-a' ($1 \leq l \leq n$; $1 \leq a \leq n$; $l > a$) to the virtual consequence 'j' ($1 \leq j \leq n$), at level 'l' equals to:

$$^{vj}A_{vx}^{l-a} = \frac{R_{vx}^{l-a}}{R_{vj}^{l}} = \frac{R_{vx}^{l-a}}{\sum_{i=1}^{i=n} R_{vi}^{l-a}} \qquad \text{Equation 6}$$

Where:

| | |
|---|---|
| $R_{vx}^{l-a}$ | Is a risk of the lower level virtual consequence '*x*' ($1 \leq x \leq n$), at level 'l-*a*' whose contribution is calculated. |
| $R_{vj}^{l}$ | Is a risk of the virtual consequence '*j*' at level '*l*' analysed. |

## Apportionment of a virtual consequence to "source" consequences

The "Source" consequences are consequences arising directly from the model before any grouping has taken a place.

If $R_{vj}^{l}$ is a risk of the virtual consequence 'j' at any level 'l' and ${}^{vj}R_{si}^{cek}$ is a risk of any "source" consequence 'i' (1 ≤ i ≤ n), arising from any Critical Event 'k' (1 ≤ k ≤ n) and feeding the virtual consequence 'j', then,

$$R_{vj}^{l} = \sum_{k=1}^{k=n} \sum_{i=1}^{i=n} {}^{vj}R_{si}^{cek}$$

Equation 7

Where:

$\sum_{k=1}^{k=n} \sum_{i=1}^{i=n} {}^{vj}R_{si}^{cek}$ Is a total of risk of all source consequences '*i*' arising from different Critical Events '*k*'

Hence, the fractional contribution of the "source" consequence 'y', arising from the Critical Event 'h' and feeding the virtual consequence 'j', to the virtual consequence 'j' at any level 'l' equals to:

$${}^{vj}A_{sy} = \frac{{}^{vj}R_{sy}^{ceh}}{R_{vj}^{l}} = \frac{{}^{vj}R_{sy}^{ceh}}{\sum_{k=1}^{k=n} \sum_{i=1}^{i=n} {}^{vj}R_{si}^{cek}}$$

Equation 8

Where:

${}^{vj}R_{sy}^{ceh}$ Is the risk of a "source" consequence '*y*' whose contribution is calculated.

$R_{vj}^{l}$ Is a risk of the virtual consequence '*j*' at any level '*l*' analysed.

## Apportionment of a virtual consequence to Core Hazards (Critical Events)

The risk arising from the Core Hazard (Critical Event) is equal to a total of risks embodied within all the "source" consequences arising from the Core Hazard. However, contribution of a Core Hazard to a Virtual Consequence equates to a sum of the risks embodied within only those "source" consequences originating from the Core Hazard which are grouped into the Virtual Consequence analysed. Therefore vjAceh, fractional contribution of the Critical Event 'h' ($1 \le h \le n$) to the virtual consequence 'j' at level 'l' and thus equates to:

$$ {}^{vj}A_{ceh} = \frac{\sum_{i=1}^{i=n} {}^{vj}R_{si}^{ceh}}{\sum_{k=1}^{k=n}\sum_{i=1}^{i=n} {}^{vj}R_{si}^{cek}} \qquad \text{Equation 9} $$

Where:

$\sum_{i=1}^{i=n} {}^{vj}R_{si}^{ceh}$ Is the sum of risks of all the "source" consequences '*i*' arising from the critical event '*h*' feeding into a virtual consequence '*j*'.

$\sum_{k=1}^{k=n}\sum_{i=1}^{i=n} {}^{vj}R_{si}^{cek}$ Is the sum of risks of all the "source" consequences '*i*' arising from all critical events '*k*' ($1 \le k \le n$) feeding into a virtual consequence '*j*'.

## Importance of Base Events relative to a "source" consequence or a virtual consequence

In order to define the contributions from base events to the consequences, the decision was made to use the Fussel-Vesely (FV) importance value. Fussel-Vesely importance is a measure of the contribution that a constituent of the fault tree makes to the top event, probability, or frequency. It is defined as the sum of the probabilities or frequencies of all cutsets in which the constituent appears, divided by the top event probability or frequency. This value is sometimes referred to as Fractional Importance. In effect, the Fussel-Vesely importance is a measure of the sensitivity of the top event to changes in the base event.

Fractional importance of the Base Event 'x' ($1 \le x \le n$) to the Critical Event (Hazard) 'k' regarding Fussel-Vesely value is then given by:

$$^{cek}I_{bex} = \frac{FV_{bex}}{\sum_{i=1}^{i=n} FV_{bei}}$$

Equation 10

Where:

$FV_{bex}$ Is the Fussel-Vesely importance value for the Base Event '$x$'.

$\sum_{i=1}^{i=n} FV_{bei}$ Is the total Fussel Vesely importance value of the model (sum of FV values of all Base Events '$i$' within the model).

The fractional importance of the Base Event 'x' ($1 \le x \le n$) to the "source" consequence 'y', originating from Critical Event 'h', regarding Fussel-Vesely value is same as the Fractional importance of the Base Event 'x' ($1 \le x \le n$) to the Critical Event (Hazard) 'h' giving a rise to the "source" consequence 'y' and is given by:

$$^{sy}I^{ceh}_{bex} = {}^{ceh}I_{bex} = \frac{FV_{bex}}{\sum_{i=1}^{i=n} FV_{bei}}$$

Equation 11

Where:

$FV_{bex}$ Is the Fussel Vesely importance value for the Base Event '$x$'.

$\sum_{i=1}^{i=n} FV_{bei}$ Is the total Fussel Vesely importance value of the model (sum of FV values of all Base Events '$i$' within the model).

Fractional importance of the Base Event 'x' ($1 \le x \le n$) to the virtual consequence 'j' at any level 'l', regarding Fussel-Vesely value, is relative to the fractional contribution of the "source" consequence 'y' ($1 \le y \le n$), originating from Critical Event 'h', the base event 'x' is causing, to the virtual consequence 'j' at any level 'l'. Therefore:

$$ {}^{vj}I_{bex} = {}^{sy}I^{ceh}_{bex} \cdot {}^{vj}A^{ceh}_{sy} = \frac{FV_{bex}}{\sum_{i=1}^{i=n} FV_{bei}} \cdot \frac{{}^{vj}R^{ceh}_{sy}}{R^{l}_{vj}} = \frac{FV_{bex}}{\sum_{i=1}^{i=n} FV_{bei}} \cdot \frac{{}^{vj}R^{ceh}_{sy}}{\sum_{k=1}^{k=n}\sum_{i=1}^{i=n} {}^{vj}R^{cek}_{si}} $$

Equation 12

Where:

| | |
|---|---|
| ${}^{vj}I_{bex}$ | Is the Fractional importance of the Base Event '*x*' (1 ≤ *x* ≤ n) to the virtual consequence '*j*'. |
| ${}^{sy}I^{ceh}_{bex}$ | Is the Fractional importance of the Base Event '*x*' (1 ≤ *x* ≤ n), causing a Critical Event '*h*' to the "source" consequence '*y*'. |
| ${}^{vj}A^{ceh}_{sy}$ | Is the fractional contribution of the "source" consequence '*y*' (1 ≤ *y* ≤ n), arising from the Critical Event '*h*', to the virtual consequence '*j*' at any level '*l*'. |
| $FV_{bex}$ | Is the Fussel-Vesely importance value for the Base Event '*x*'. |
| $\sum_{i=1}^{i=n} FV_{bei}$ | Is the total Fussel-Vesely importance value of the model (sum of FV values of all Base Events '*i*' within the model). |
| ${}^{vj}R^{ceh}_{sy}$ | Is risk of the "source" consequence '*y*', arising from the Critical Event '*h*' |
| $R^{l}_{vj}$ | Is a risk of the virtual consequence '*j*' at any level '*l*' analysed. |
| $\sum_{k=1}^{k=n}\sum_{i=1}^{i=n} {}^{vj}R^{cek}_{si}$ | Is the sum of risks of all the "source" consequences '*i*' arising from all critical events '*k*' (1 ≤ *k* ≤ n) feeding into a virtual consequence '*j*'. |

## Importance of Barriers and Splitters relative to a "source" consequence or a virtual consequence

A measure of apportionment (importance) of the Barrier or Splitter is a measure of its "capability" to convey the frequency or probability of consequence occurrence towards the lower loss consequence.

For any consequence model element 'j ' ( $\le j \le$ [illegible] frequency at the consequence model element, fj, equals the sum of all the frequencies feeding into the consequence model element 'j'. The frequency coming out of the consequence model element is defined as a product of the frequency at the consequence model element and the probability of the consequence model element output.

Note: The Consequence model element can be a barrier, a splitter or a critical event.

Therefore, frequency at the consequence model element ‘ g ’ is:

$$f_\gamma = \sum_{j=\alpha}^{j=\zeta} \sum_{i=1}^{i=2} {}^{\gamma}p_i^j \cdot f_j$$ Equation 13

Where:

| | |
|---|---|
| $f_\gamma$ | Is the frequency at the consequence model element ‘ γ ’. |
| ${}^{x}p_i^j$ | Is the probability of any output ‘*i*’ of any consequence model element ‘*j*’ feeding into the consequence model element ‘γ’. |
| $f_j$ | Is the frequency at any consequence model element ‘*j* ’, whose output ‘*i*’ is feeding into the consequence model element ‘γ’. |

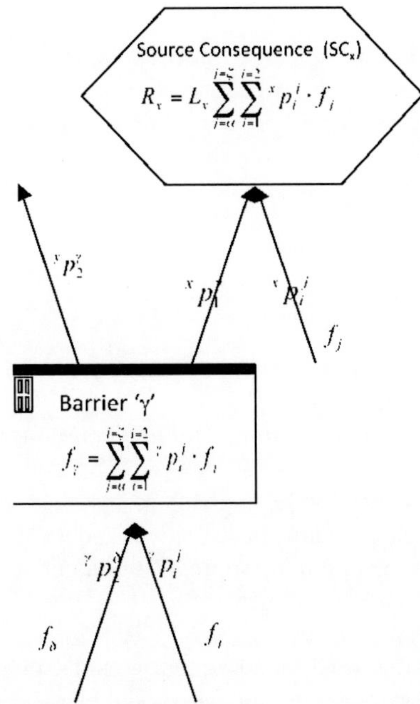


Figure B-1: Consequence model notation

It follows that the Risk of a "source" Consequence 'x ' is a product of the Loss of the consequence and the sum of all the frequencies feeding into the consequence. Therefore:

$$R_x = L_x \sum_{j=\alpha}^{j=\zeta} \sum_{i=1}^{i=2} {}^{\gamma}p_i^j \cdot f_j \qquad \text{Equation 14}$$

Where:

$R_x$ Is a risk of the "source" consequence '*x* '.

$L_x$ Is a loss of the "source" consequence '*x* '.

${}^{x}p_i^j$ Is probability of any output '*i*' of any consequence model element '*j* ' feeding into the "source" consequence '*x* '.

$f_j$ Is the frequency at any consequence model element '*j* ', whose output '*i*' is feeding into "source" consequence '*x* ' analysed.

The 'weight' of a consequence model element's output or risk, conveyed through an output 'i' of a model element 'j ', to a "source" consequence 'x ' or the next consequence model element 'x ' is equal to:

$$Wr_i^j = L_x \cdot {}^{\gamma}p_i^j \cdot f_j \qquad \text{Equation 15}$$

Where:

$L_x$ Is a loss of the "source" consequence '*x* ' or the next consequence model element '*x*'.

${}^{x}p_i^j$ Is probability of any output '*i*' of any consequence model element '*j* ' feeding into the "source" consequence '*x*' or the next consequence model element '*x* '.

$f_j$ Is the frequency at any consequence model element '*j* ', whose output '*i*' is feeding into "source" consequence '*x* ' or the next consequence model element '*x* '.

It follows that the risk at the consequence model element equals the sum of risks conveyed from a consequence model element's outputs and is equal to:

$$R_{MEj} = \sum_{i=1}^{i=5} Wr_i^j$$

Equation 16

Hence, loss at the model element 'j ' equals to the risk at the consequence model element divided by frequency at the consequence model element:

$$L_{MEj} = \frac{R_{MEj}}{f_j}$$ Equation 17

Absolute importance of the consequence model element 'j ', originating from Critical Event 'h' is equal to the maximum difference between outputs' weights (risk conveyed through all outputs 'i' of a consequence model element 'j ') to a "source" consequence 'x ' or the next consequence model element 'x ' and is expressed as:

$$_{absolute}I_{MEj}^{ceh} = \max\left| Wr_i^j - Wr_{i-1}^j \right|_{i=2}^{i=5}$$ Equation 18

Where:

| | |
|---|---|
| $_{absolute}I_{MEj}^{ceh}$ | Is an absolute importance of the consequence model element '*j* '. |
| $Wr_i^j$ | Is the weight of the output '*i*' of a consequence model element '*j* '. |
| $Wr_{i-1}^j$ | Is the weight of the output '*i-1*' of a consequence model element '*j* '. |
| $\max\left\| Wr_i^j - Wr_{i-1}^j \right\|_{i=2}^{i=5}$ | Is the maximum value of the difference in weight between all outputs '*i*' of a consequence model element '*j* '. |

The fractional importance of the consequence model element 'w ' (1 ≤ w ≤ n), originating from Critical Event 'h', relative to the whole consequence model structure is given by:

$$I^{ceh}_{MEj} = \frac{{}_{absolute}I^{ceh}_{ME\omega}}{\sum_{j=1}^{j=n} {}_{absolute}I^{ceh}_{MEj}} \qquad \text{Equation 19}$$

Where:

| | |
|---|---|
| ${}_{absolute}I^{ceh}_{ME\omega}$ | Is an absolute importance of the consequence model element '$\omega$', originating from Critical Event '$h$'. |
| $\sum_{j=1}^{j=n} {}_{absolute}I^{ceh}_{MEj}$ | Is the sum of all absolute importance values of all the consequence model elements '$j$', originating from Critical Event '$h$'. |

The fractional importance of the consequence model element 'w ' (1 ≤ w ≤ n), to the "source" consequence 'y', originating from Critical Event 'h' is given by:

$${}^{scy}I^{ceh}_{ME\omega} = \frac{{}^{scy}_{absolute}I^{ceh}_{ME\omega}}{\sum_{j=1}^{j=n} {}^{scy}_{absolute}I^{ceh}_{MEj}} \qquad \text{Equation 20}$$

Where:

| | |
|---|---|
| ${}^{scy}_{absolute}I^{ceh}_{ME\omega}$ | Is an absolute importance of the consequence model element '$\omega$', originating from Critical Event '$h$' and on the path to the "source" consequence '$y$'. |
| $\sum_{j=1}^{j=n} {}^{scy}_{absolute}I^{ceh}_{MEj}$ | Is the sum of all absolute importance values of all the consequence model elements '$j$', originating from Critical Event '$h$' and on the path to the "source" consequence '$y$'. |

The fractional importance of the consequence model element 'w ' (1 ≤ w ≤ n), on the path to the "source" consequence 'y', to the virtual consequence 'j', at any level 'l', is relative to the fractional contribution of the "source" consequence 'y' (1 ≤ y ≤ n), originating from Critical Event 'h' to the virtual consequence 'j', and is given by:

$$ {}^{vj}I^{ceh}_{ME\omega} = {}^{vj}A_{sy} \cdot {}^{scy}I^{ceh}_{ME\omega} = \frac{{}^{vj}R^{ceh}_{sy}}{\sum_{k=1}^{k=n}\sum_{i=1}^{i=n} {}^{vj}R^{cek}_{si}} \cdot \frac{{}^{scy}_{absolute}I^{ceh}_{ME\omega}}{\sum_{j=1}^{j=n} {}^{scy}_{absolute}I^{ceh}_{MEj}} \qquad \text{Equation 21} $$

Where:

${}^{scy}_{absolute}I^{ceh}_{ME\omega}$ Is an absolute importance of the consequence model element '$\omega$', originating from Critical Event '$h$' and on the path to the "source" consequence '$y$'.

$\sum_{j=1}^{j=n} {}^{scy}_{absolute}I^{ceh}_{MEj}$ Is the sum of all absolute importance values of all the consequence model elements '$j$', originating from Critical Event '$h$' and on the path to the "source" consequence '$y$'.

${}^{vj}R^{ceh}_{sy}$ Is the risk of the "source" consequence '$y$' on whose path the consequence model element '$\omega$' is.

$R^{l}_{vj}$ Is the risk of the virtual consequence '$j$' at any Layer '$l$' analysed.

# APPENDIX B

# SPECIFICATION FOR CAUSE-CONSEQUENCE MODELS INTEGRATION ENVIRONMENT

<table>
<tr><td colspan="2">Object Type/Class</td></tr>
<tr><td colspan="2">Critical Event Model super object (real consequences and base events of the model)</td></tr>
<tr><td>Object Attributes</td><td>Each of the Cause-Consequence models within the ISAE is presented through two distinguished sets of records:<br>1. A record set providing basic information about all the consequences arising from the model (consequence name and reference number);<br>2. Basic information about all the base events from the model (consequence name and reference number).<br>Visual appearance is provided below. Real Consequences cannot receive any frequencies from outside models. Real Consequences can only be fed into the base events of other models, intermittent operators or into high level/virtual consequences. Multi-coloured circles present linking points:<br>1. Real consequences linked to virtual consequences have red circles associated with them;<br>2. Real consequences linked to intermittent operator have green circles associated with them;<br>3. Real consequences linked to base events of other models have blue circles associated with them, and<br>4. Critical events linked to base events of other models have orange circles associated with them<br>5. Real consequences that are disabled (not to be taken into account for calculations of total risk, but still calculated for individual models risk) are to be notified by grey circles. Real consequences can be disabled manually, through a control switch or if the critical event has been linked as the feeder to the base event of some other model.<br>Base events can be linked to other models only as recipients of frequencies coming out of consequences or critical events calculated within external models or intermittent operator linked to consequences.</td></tr>
</table>

| Object Type/Class | |
|---|---|
| Critical Event Model super object (real consequences and base events of the model) | |
| | Base events can be fed by real consequences or critical events of other models. Base events related linking points are presented by collared circles as follows:<br>1. Base events fed by real consequences/critical events of other models are presented by blue circles;<br>2. Base events that are independent from other models are presented by grey circles;<br>3. In case of common base events (same base events recurring throughout several different models, but not necessarily with the same frequency) these are notified by green circles;<br>4. Base events determining the probability of the barrier failure are notified by black circles. Values for these base events are expressed as probabilities only. |
| Object Referencing | References of Cause-Consequence models are the same as the references of the Critical Events that the models are built around. References of real consequences, critical events and base events are the same as the references within the model. |
| Object Representation | 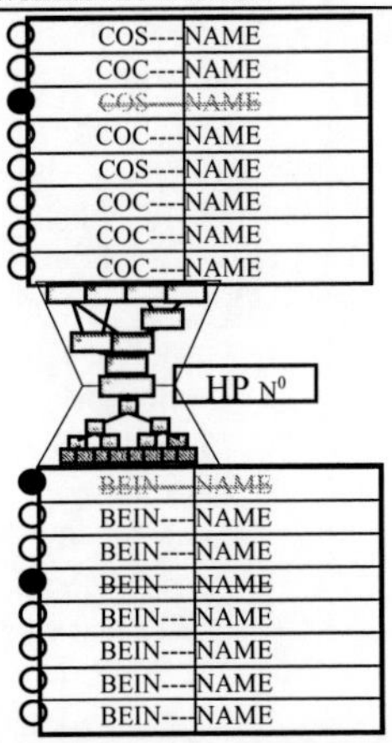 COS---NAME<br>COC---NAME<br>COS---NAME<br>COC---NAME<br>COS---NAME<br>COC---NAME<br>COC---NAME<br>COC---NAME<br>HP N⁰<br>BEIN---NAME<br>BEIN---NAME<br>BEIN---NAME<br>BEIN---NAME<br>BEIN---NAME<br>BEIN---NAME<br>BEIN---NAME<br>BEIN---NAME |
| Data referencing | 1. Data transferred from a real consequence to the base event is referenced as follows: FROM “consequence reference” TO “base event reference”.<br>2. Data transferred from a critical event to the base event is referenced as follows: FROM “critical event reference” TO “base event reference”;<br>3. Data transferred from the real consequence to the virtual consequence is referenced as follows: FROM “consequence reference” TO “virtual consequence reference”; |

| Object Type/Class | |
|---|---|
| Critical Event Model super object (real consequences and base events of the model) | |
| | 4. Data transferred from the real consequence to the *intermittent operator* is referenced as follows: FROM "consequence reference" TO "*intermittent operator* reference";<br>5. Data transferred from a critical event to the base event is referenced as follows: FROM "critical event reference" TO "base event reference";<br>6. Data transferred from the critical event to the *intermittent operator* is referenced as follows: FROM "critical event reference" TO "*intermittent operator* reference". |
| Object values and default conditions/data | Default values for base events are values provided originally within the model.<br>If the base event is linked with another model's consequence or critical event then the value of the base event is the frequency of the consequence/critical event calculated by the other model.<br>A default value of virtual consequences is null for all of the loss category's risks of the consequence. Real consequences feeding the virtual consequence are passing risks calculated by individual models for all the loss categories to the virtual consequence. The value of risk for each loss category within the virtual consequence is calculated as a sum of all the risks passed to the virtual consequence from all the real consequences feeding the virtual consequence. |
| Combinational Rules | 1. Only frequencies are passed from real consequences to base events;<br>2. Linked base events can not be in an AND relationship with other base events with frequencies used as units within the model structure. If this is the case then a warning message must be displayed. It is possible that frequency feeding the base event can be translated into probability using the *TRANSFER FUNCTION* object;<br>3. Risk calculated for each loss category are passed from real consequences to virtual consequences;<br>4. Aggregated risks are passed from a virtual consequence to virtual consequence;<br>5. It is ensured that no closed loops are formed within the integrated model;<br>6. Frequencies are passed from critical events to base events if the base event is determining the frequency of the critical event;<br>7. If the base events defined by the other model are within the fault trees that determine the probability of the barrier |

| Object Type/Class | |
|---|---|
| Critical Event Model super object (real consequences and base events of the model) | |
| | failure, it must be ensured that the total outcome of such a fault tree is expressed as a probability;<br>8. All the rules applicable within the existing ISAE should still apply. |
| Working Environment of the object | Model Integration Environment |

| Object Type/Class | |
|---|---|
| Virtual Consequences | |
| Object Attributes | The value of these objects is calculated as the sum of risks for each risk category for all the feeders into the virtual consequence. |
| Object Representation | VRCC---- VREC---- VRSC---- VRBC---- |
| Default conditions/data | Default value is null. |
| Working Environment of the object | Model Integration Environment |

| Object Type/Class | |
|---|---|
| Inter-model IN and OUT linking objects/symbols | |
| Object Attributes | These two objects are used for linking between two models during the modelling, from within one work sheet to another. "In From: - - - " object feeds into the selected base event. "Output to: - - - " object feeds into selected base event of another model, selected virtual consequence or *intermittent operator*. These two objects are in correspondence with links established through Critical Event Model super object, i.e. all the links between models created within the Model Integration Environment are automatically mimicked by an IN and OUT symbols, automatically created within the corresponding models worksheets. The inverse applies as well, i.e. links defined through the models worksheets, using IN and OUT symbols are mimicked by the Model Integration Environment. Visual appearance is provided below. |

| Object Referencing | Objects are referenced as follows:<br>1. Out to virtual consequence symbol: "Out to: *virtual consequence reference*";<br>2. Out to base event symbol: "Out to: *base event/critical event (other model) reference*";<br>3. In from symbol: "In from: *consequence/critical event (other model) reference*" or "In from: *intermittent operator reference*". |
|---|---|
| Object Representation | OUT symbols: Real consequences linked to virtual consequences have red arrow associated with them and real consequences or critical events linked to base events of other models have blue arrow associated with them. Real consequences are linked to an OUT symbol by a blue or red line.<br>IN symbols: IN symbol have grey arrow associated with them. In Symbol is linked to base event by a grey line. <br>In from: HP700    Out to: HP700    Out to: C ----- |
| Data referencing | Data transferred from a real consequence to the base event is referenced in the following way: FROM "consequence reference/critical event" TO "base event reference".<br>Data transferred from the real consequence to the virtual consequence is referenced as: FROM "consequence reference" to "virtual consequence reference" |
| Object values | Objects take values as follows:<br>1. Out to virtual consequence symbol: risks calculated by individual models for all the loss categories;<br>2. Out to base event symbol: the frequency of the consequence calculated by the model;<br>3. In from any symbol: risks calculated by individual models for all the loss categories or the frequency of the consequence (link from) calculated by the other model;<br>4. Default values for base events are values provided originally within the model. If the base event is linked with other model's consequence than the value of the base event;<br>5. Default value of virtual consequences is null for all the loss categories risks of the consequence. Value of risk for each loss category within the virtual consequence is calculated as a sum of all the risks passed to the virtual consequence from all the real consequences feeding the virtual consequence. |

| | |
|---|---|
| Combinational Rules | 1. IN symbol can fed in Base Event only;<br>2. Base events can be linked to other models only as recipients of frequencies coming out of consequences/critical events calculated within external models;<br>3. Real consequences/critical events cannot receive any direct input from outside models. Real consequences can only fed into base events of other models or into high level/virtual consequences;<br>4. Critical event can only fed into base events of other models;<br>5. If the base events defined by the other model are within the fault trees that determine the probability of the barrier, it must be ensured that the total outcome of such a fault tree is expressed as a probability (self check must operate on the structure instantaneously). |
| Object controls/data fields | Fields providing the information about the values passed through the link. In case of an IN symbol, i.e. link from a real consequence to the base event, frequency passed is displayed within the symbol's dialog box. In case of an OUT symbol two possibilities exist:<br>1. If the link is between the real consequence/critical event and the base event information displayed within the symbol's dialog box should be the frequency passed through the link;<br>2. If the link is between the two consequences (any consequences) the information passed through the link shall be risk calculated for all the existing risk categories. |
| Working Environment of the object | Individual models worksheets |

| Object Type/Class | |
|---|---|
| Supper Connectors *intermittent operators* | |
| Object Attributes | Unlimited number of inputs into this object and only one output. The output of this object is the sum of all the inputs. |
| Object Referencing | Same/similar as the super-connector within the existing individual models worksheets environment. |
| Object Representation | |

<table>
<tr><td>Object values</td><td colspan="2">The output of the object is the sum of all the inputs. All the inputs must be of the same type i.e. either frequencies, probabilities or risks. If inputs are frequencies/ probabilities the output is the sum of all the frequencies/ probabilities. If inputs are risks then the output is the sum of all the inputs, but aggregation must be done individually for each risk category. Consequently, the output in this case will be in form of 13 values, one for each risk category.</td></tr>
<tr><td>Combinational Rules</td><td colspan="2">All the inputs must be of the same type i.e. either frequencies or risks.</td></tr>
<tr><td colspan="2">Working Environment of the object</td><td>Model Integration Environment</td></tr>
</table>

<table>
<tr><td colspan="2">Object Type/Class</td></tr>
<tr><td colspan="2">Transfer function intermittent operator</td></tr>
<tr><td>Object Attributes</td><td>This object can have multiple inputs and outputs. Inputs are processed according to the “transfer function” defined within the object to yield outputs. For the transfer function it is possible to define inputs and outputs as the parameters of the “transfer function”. Inputs may come from real consequences or high level/virtual consequences only or critical events. Two types of inputs may exist:<br>1. If inputs are risks passed through, from consequences, the object must support distinction between different categories of risks. Therefore, it is possible to define each of the risk categories as a separate parameter as well as to lump them all together if needed. In this case outputs must be risks as well;<br>2. If inputs are frequencies/probabilities passed form the consequences/critical events to the object the object shall support a simple allocation of the input to the parameter. In this case outputs may be either frequencies or risks depending on the nature of the transfer function.<br>Four types of parameters may be used:<br>1. Parameters used/defined to identify inputs;<br>2. Parameters used/defined to identify outputs;<br>3. Parameters defined to establish/build the transform function;<br>4. Parameters already existing within the database and used to establish/build the transform function.<br>Immediate calculation of the “transform function” shall be available allowing for the modeller to check the function and results instantly.</td></tr>
</table>

| | |
|---|---|
| Object Referencing | Object is referenced as TRFnnn, where TRF is a prefix and nnn is three digit identification. |
| Object Representation | TRFnnn |
| Data referencing | 1. Each of the inputs is referenced separately as an input to the object;<br>2. Each of the outputs is referenced separately as an output of the object;<br>3. The transfer function is referenced to the object;<br>4. All the additional information such as "rationale, justification, etc. are referenced to the object;<br>5. All the parameters defined are referenced individually, or if parameters already defined elsewhere are to be used then reference to these parameters relative to the object must be established. |
| Object values | Calculated or defaulted values. |
| Default conditions/data | Transfer function is "null", i.e. inputs to the object are not linked to the outputs of the object (outputs of the object are set to zero). |
| Combinational Rules | 1. All the inputs into the object must be of the same units, either frequencies or risks;<br>2. Combination of inputs defined by the transfer function yields outputs defined by the transfer function;<br>3. If the output is to be linked to the base event then the unit of the output can be frequency or probability depending on the causal model structure the base event is positioned within;<br>4. If the output is to be linked to the consequence then the unit of the output must be the risk calculated for all the risk categories. |
| Working Environment of the object | Model Integration Environment |

| Object Type/Class | |
|---|---|
| Links from real/ virtual consequence to virtual consequence | |
| Object Attributes | Only risks calculated for each risk category shall be passed by the link. |
| Object Representation | |
| Combinational Rules | Link from the real/ virtual consequence to the virtual consequence or any of the intermittent operators |
| Working Environment of the object | Model Integration Environment |

| Object Type/Class | |
|---|---|
| Splitter *intermittent operator* | |
| Object Attributes | Two types of inputs may exist:<br>1. If inputs are risks passed through from consequences, the object supports distinction between different categories of risks. Therefore, it is possible to define each of the risk categories as a separate input as well as to lump them all together if needed. Split is either per individual risk category or for the total/lumped value. Outputs must be risks as well;<br>2. If inputs are frequencies/probabilities passed form the consequences to the object, the object supports a simple allocation of the input to the parameter. Outputs must be frequencies.<br>Other attributes are same as for the splitter already existing within the worksheet environment. |
| Object Representation | 25% 75% Splitter |
| Working Environment of the object | Model Integration Environment |

| Object Type/Class | |
|---|---|
| Links from the critical event to the base event | |
| Object Attributes | Only frequency calculated for the critical event shall be passed through the link. |
| Object Representation | |
| Object values | Frequency calculated for the critical event |
| Combinational Rules | Links from the critical event to the base event or any of the intermittent operators. |
| Working Environment of the object | Model Integration Environment |

| Object Type/Class | |
|---|---|
| Logical Gates (AND, OR) *intermittent operators* | |
| Object Attributes | The algorithm for calculations of gates is provided within the causal domain specification.<br>Only links from real consequence or critical events to the base event can be linked to the gate.<br>In case of an AND gate, all the frequencies feeding the gate except for one, must be transferred into probabilities within the Transfer function object. Links from real consequence/virtual consequence to virtual consequence cannot be linked into gates. |
| Object Representation | OR & |
| Combinational Rules | Only Links from real consequence/critical event to base events can be linked to the gate.<br>No more than one frequency can fed the AND gate. |
| Object Controls/ linking and processing | Linking of inputs into the gate. Linking of output from the gate. Processing (same as described within the causal analysis tool specification). |
| Working Environment of the object | Model Integration Environment |

| Object Type/Class | |
|---|---|
| Links from real consequence to base event | |
| Object Attributes | Only frequency calculated for the consequence shall be passed through the link. |
| Object Referencing | To be defined |
| Object Representation | |
| Data referencing | To be defined |
| Object values | Frequency calculated for the consequence |
| Combinational Rules | Links from the real consequence to the base event or any of the intermittent operators. |
| Working Environment of the object | Model Integration Environment |

# REFERENCES AND BIBLIOGRAPHY


(ATKINS, 2002) ATKINS GLOBAL (2002) *Axle Counter Concept Safety Case*, London: ATKINS GLOBAL (Reference 1956007-L-SC-001, v7.00, dated 22-03-2002).

(ATKINS, 2003) ATKINS GLOBAL (2003) *Systems Integration*; London: ATKINS GLOBAL (Reference 5012008-51-001 Issue 1; 28/11/2003).

(Bernstein, 1996) Bernstein P. L. (1996) *Against the Gods: Remarkable Story of Risk.* USA: John Wiley & Sons, INC; ISBN-471-29563-9.

(Bhatnagar and Kanal, 1991) Bhatnagar R. and Kanal N. L. (1991) *Models of enquiry and formalism for approximate reasoning*; USA: University of Maryland College Park.

(Blanchard and Fabrycky, 1998) Benjamin S. Blanchard and Wolter J. Fabrycky (1998) *Systems Engineering and Analysis.* New Jersey: Prentice-Hall, Inc. ISBN 0-13-135047-1.

(Bobbio and Portinali, 1999) Andrea Bobbio and Luigi Portinali (1999) Comparing different methodologies of Probabilistic Structured Modelling by the analysis of typical industrial dependable systems. *CENELEC WGA10 Workshop: Bridging the Gap to Railway Interoperability*, Germany, Munich.

(BS EN 50126: 1999) BRITISH STANDARDS INSTITUTE (1999) *Railway Applications – The specification and demonstration of dependability, reliability, availability, maintainability and safety (RAMS)* London: British Standards Institute.

| | |
|---|---|
| (BS EN 50129, 2003) | BRITISH STANDARDS INSTITUTE (2003) *Railway Applications: Software for Railway Control and Protection Systems*. London: British Standards Institute. ISBN 0 580 41814 6. |
| (BS EN 61508: 2003) | BRITISH STANDARDS INSTITUTE (2003) *Functional Safety of electrical/electronic/programmable electronic safety-related systems* London: British Standards Institute. |
| (CENELEC 1998) | CENELEC (1998) *TC9XA-WGA10- A Method for SIL Allocation*. EU: CENELEC. |
| (CENELEC 2003) | CENELEC EN (2003) *50129: Railway applications - Communication, signalling and processing systems - Safety related electronic systems for signalling.* EU: CENELEC. |
| (Checkland, 1984) | Checkland P. (1984) *Systems Thinking Systems Practice*. England: John Wiley & Sons Ltd; ISBN 0-471-27911-0. |
| (Civil Aviation Authority, 2006) | CIVIL AVIATION AUTHORITY (2006) *CAP 760 Guidance on the Conduct of Hazard Identification, Risk Assessment and the Production of Safety Cases For Aerodrome Operators and Air Traffic Service Providers*. UK: Safety Regulation Group. |
| (Fenton and Hill, 1993) | Fenton N. and Hill G. (1993) *Systems Construction and Analysis, A Mathematical and Logical Framework.* London: McGraw-Hill Book Company Europe, ISBN 0-07-707431-9. |
| (Fenton, 1991) | Fenton N. (1991) *Software Metrics – A Rigorous Approach*. London: Chapman and Hall, ISBN 0-412-40440-0. |
| (Fenton, Neil and Forey, 2001) | Fenton N. , Neil M., Forey S. and Harris R. (2001) Using BBNs to predict the reliability of military vehicles. *Computing and control engineering journal*; pp 11-20. |

(Fragola and Spahn, 1973) Fragola, J.R. and Spahn J.F. (1973), The Software Error Effects Analysis; A Qualitative Design Tool, In: *Proceedings of the 1973 IEEE Symposium on Computer Software Reliability*, IEEE, New York, pp. 90-93

(Goble, 1998) Goble W. M. (1998) *The use and development of quantitative reliability and safety analysis in new product design*, PhD), Technishe Universiteit Eindhoven, ISBN 60-386-0870-5.

(Health and Safety Executive, 2001) GREAT BRITAIN. HEALTH AND SAFETY EXECUTIVE (2001) *Reducing risks, protecting people*. London: Her Majesty's Stationery Office ISBN 0717621510

(Health & Safety Commission, 1998) GREAT BRITAIN. HEALTH AND SAFETY COMMISSION (1998) *Rail Safety: Proposal for Regulations on train protection systems and mark 1 stock (Consultative Document)*. London. HSE Books.

(Hessami 2004) Hessami, A., (2004) A systems framework for Safety and Security: the holistic paradigm. *Systems Engineering*; Volume 7; ISSN 1098-1241.

(Hessami and Hunter, 2002) Hessami A. and Hunter A. (2002) Formalization of weighted factors analysis, *Knowledge-Based Systems*, Issue 15, pp. 377-390.

(Hessami, 1999a) Hessami, A. (1999) Risk, a Missed Opportunity. *Risk & Continuity-The International Journal for Best Practice Management*, Volume 2, Issue2.

(Hessami, 1999b) Dr Ali Hessami, (1999) Risk Management: A Systems Paradigm. *Systems Engineering*, Volume 2, Number 3.

(Hsu and Musicki, 2005) Feng Hsu and Zoran Musicki (2005) Issues and Insights of Probabilistic Risk Assessment Methodology in Nuclear and Space Applications. In: *Proceedings of the IEEE International Conference on Systems, Man and Cybernetics*, ISBN: 0-7803-9298-1.

| | |
|---|---|
| (Karcanias and Hessami, 2009) | Karcanias N. and Hessami A. (2008) *Complexity and the Notion of System of Systems: Part (I): General Systems and Complexity*; Unpublished paper, City University, London. |
| (Karcanias and Hessami, 2009) | Karcanias N. and Hessami A. (2008) *Complexity and the Notion of System of Systems: Part (II): Defining the notion of System of systems*; Unpublished paper, City University, London. |
| (Karcanias, 2003) | Karcanias N (2003) *System Concepts for General Processes: Specification of a New Framework*. Unpublished paper, City University, London. |
| (Karcanias, 2004) | Karcanias N (2004) Complex Systems: Modelling and Simulation Challenges. *SIM-SERV workshop: Working Group on Modelling and Simulation* Greece, University of Patras. |
| (Kelly and Weaver, 2004) | Kelly T. and Weaver R. (2004) The Goal Structuring Notation – A Safety Argument Notation. In: *Proceedings of Dependable Systems and Networks 2004 Workshop on Assurance Cases*. UK. |
| (Kelly, 2001) | Contract Research Report for QinetiQ COMSA/2001/1/, Kelly, T. (2001) *Concepts and Principles of Compositional Safety Case Construction*. York: Department of Computer Science, University of York. |
| (Kirwan, 1996) | Barry Kirwan (1996) The validation of 3 Human Reliability Quantification techniques – THERP, Heart and JHDI, *Applied Ergonomics* Vol. 27. No 6pp.359-373. |
| (Linstone and Turoff , 2002) | Linstone H. A. and Turoff M.(2002) *The Delphi Method: Techniques and Applications*. New Jersey, New Jersey Institute of Technology. |

(Lucic and Short, 2007) Lucic I. and Short R. (2007) Engineering Safety and Assurance Case. In: *Proceedings of the Institution of Engineering and Technology Seminar on Safety Assurance, 7 – 8 November 2007*. London.

(Lucic, 2001) ATKINS RAIL REPORT- Lucic I. (2001) *Proposal for development of methodology for risk based application of TPWS to speed restriction, Version 01*. London.

(Lucic, 2002) ATKINS RAIL REPORT- Lucic I. (2002) *Manchester South Capacity Improvement Project: Hazard Forum Procedure (4026002-L-RM-004; Version 3.00; 25/11/2002)*. London.

(Lucic, 2003a) ATKINS RAIL REPORT- Lucic I. (2003) *Axle Counter Risk Analysis* (Reference ET-JWR-010033541). London.

(Lucic, 2003b) ATKINS RAIL REPORT- Lucic I. (2003) *Axle Counter – Modelling Enhancements, High Level Hazard Identification Report* (Reference 1956710-L-HA-001, version 00.04, dated 09-01-2003). London.

(Lucic, 2003c) ATKINS RAIL REPORT- Lucic I. (2003) *Stage 1 - Test Plan*, (Reference 195620-L-TG-001, version 1.0, dated 14/05/03). London.

(Lucic, 2003d) ATKINS RAIL REPORT- Lucic I. (2003) *Manchester South Capacity Improvement Project: Change Safety Analysis Strategy (FF05-000-SA-SAC-5204.v04.00)*. London.

(Lucic, 2004a) ATKINS RAIL REPORT- Lucic I. (2004) *Axle Counter Modelling - Stage 1 Final Report* (Reference 1956710-L-TR-001 issue 01.00). London.

(Lucic, 2004b) ATKINS RAIL REPORT- Lucic I. (2004) *Manchester South Capacity Improvement Project: Hazard Management Plan (FF05-000-SA-SAC-5004.v06.00; 24/05/2004)*. London.

(Lucic, 2005a) Lucic I. (2005) Holistic Safety Performance Forecasting For Train Detection System. In: *Proceedings of the System of Systems Engineering, IEEE-International Conference on Systems, Man and Cybernetics October 10-12, 2005.* USA, Hawaii.

(Lucic, 2005b) Lucic I. (2005-2010) Lectures "Risk Profiling: a Strategic Instrument" and "Engineering Safety and Assurance", *Delivered at: City University as part of the Risk Management module of the MSc in Energy and Environmental Technology & Economics Module*. UK, London.

(Lucic, 2005c) RAILTRACK REPORT – Lucic I. (2005) *ERTMS Hazard identification and grouping report*. London

(Lucic, 2005d) RAILTRACK REPORT – Lucic I. (2005) *ERTMS QRA model and the report*. London.

(Lucic, 2005e) RAILTRACK REPORT – Lucic I. (2005) *ISAE Enhancements specification*. London.

(Lucic, 2005f) RAILTRACK REPORT – Lucic I. (2005) *ERTMS Safety Targets – report.* London.

(Lucic, 2006) ATKINS RAIL REPORT- Lucic I. (2006) *Technical Assistance for Preparation of the Modernisation of Corridor II – Remaining Works-Development of System Description & RAMS Requirements Strategy; 24th May 2006; (Minutes of Meeting).* London.

(Lucic, 2008) Lucic I. (2008) Lecture "Engineering Safety and Assurance Case (ESAC)"; *Delivered at: The IET Course on Railway Signaling and Control Systems 2008*. UK, Birmingham.

(Lucic, 2009a) LONDON UNDERGROUND REPORT – Lucic I. (2009) *Project Management Framework Product Description and Template: Engineering Safety Management Plan (LU-PD-10655 A1).* London.

(Lucic, 2009b) LONDON UNDERGROUND REPORT – Lucic I. (2009) *Project Management Framework Product Description and Template: Engineering Safety Case (LU-PD-10656 A)*. London.

(Lucic, 2009c) LONDON UNDERGROUND REPORT – Lucic I. (2009) *Project Management Framework Product Description and Template: Engineering Safety & Assurance Case (ESAC) Case (LU-PD-10685 A1)*. London.

(Lucic, 2009d) LONDON UNDERGROUND REPORT – Lucic I. (2009) *Project Management Framework Product Description and Template: Engineering Safety Hazard Log (LU-PD-10808 A1)*. London.

(Lucic, 2009e) LONDON UNDERGROUND REPORT – Lucic I. (2009) *Project Management Framework Product Description and Template: System Safety Case (LU-PD-10739 A1)*. London.

(Lucic, 2009f) LONDON UNDERGROUND REPORT – Lucic I. (2009) *ESAC College Notes (GEN-9999-LUL-ESAC-00001)*. London.

(Ludwig, 2005) Dr. Ing. habil. Bjorn Ludwig (2005) Safety management systems-a new challenge for railways? *ETR. Eisenbahntechnische Rundschau,* vol. 53, no11, pp. 749-757. ISSN 0013-2845.

(Milloti, 2004) Dia Milloti (2004) *Systems, modelling and control in manufacturing and business processes* (PhD Transfer report) City University – London.

(Moss and Woodhouse, 1999) Moss T. R. and Woodhouse J. (1999) Criticality Analysis Revisited. *Quality and Reliability Engineering International*, Volume 15, Issue 2, pp.117-121.

(Open University Handbook, 1999) The Open University (1999) *Mathematical Methods, Models and Modelling-Handbook*. UK: OU; SUP 46655 8, 1999.

(Portland and Turoff, 2002) Harold A. Linstone Portland and Murray Turoff (2002) *The Delphi Method - Techniques and Applications*. New Jersey: New Jersey Institute of Technology; University of Southern California, ISBN 0-201-04294-0.

(Pukite and Pukite, 1998) Pukite J. and Pukite P. (1998) *Modelling for Reliability Analysis*; US: Wiley-IEEE Press, ISBN 0-7803-3482-5.

(Regan and Lucic, 2008) Regan L. and Lucic I. (2008) Use of QRA in Supporting a System Safety Case for the Victoria Line Upgrade. In: *Proceedings of the INCOSE 18th Annual International Symposium2008, System Engineering for the Planet.* The Netherlands, Utrecht.

(Reifer, 1979) Reifer D.J. (1979) Software Failure Modes and Effects Analysis. *IEEE Transactions on Reliability, Vol. R-28, No. 3.*

(Rouvroye, 2001) Rouvroye J. L. (2001) *Enhanced Markov Analysis as a method to assess safety in the process industry* (PhD), Technishe Universiteit Eindhoven, ISBN 90-386-2772-6.

(RSSB, 2007) UNITED KINGDOM. RAIL SAFETY AND STANDARDS BOARD on behalf of the UK Rail Industry (2007) *Engineering Safety Issue 3 - Yellow book.* London: RSSB; ISBN 0 9537595 0 4.

(Russo and Jain, 2001) Russo M. and Jain L.C. (2001) *Fuzzy Learning and Applications* New York: CRC Press, ISBN 0-8493-2269-3.

(Saltelli, 2002) Andrea Saltelli (2002) Sensitivity analysis for importance assessment; *Risk Analysis: an official publication of the Society for Risk Analysis.*

(SAMRAIL, 2004) European Commission Fifth Framework Programme – SAMRAIL-Consortium (2004) *Common Safety Methods, Version 3*. London: ATKINS GLOBAL.

(Short, 2007) Short R. (2007) Combining Different Types of Evidence in System Safety Assurance. In: *SARAC Asia Pacific Transport Safety & Security Conference*. China, Beijing.

(Short, 2009) Short R. (2009) The Use and Misuse of SIL; *IET/IRSE Technical Meeting*. London: IRSE.

(Sterman, 2000) Sterman J. D. (2000) *Business Dynamics – Systems Thinking and Modelling for a Complex World*. US: Irwin McGraw-Hill I. ISBN 0-07-231135-5.

(Thomas and Stupples, 2006) Thomas P. J. and Stupples D. W. (2006) J Value: a universal scale for health and safety spending. *Measurement + Control*; Vol.39/9 pp. 273-276.

(Topintzi, 2001) Topintzi E. (2001) *System Concepts and Formal Modelling Methods for Business Processes* (PhD), City University – London.

(Waldrop, 1992) Waldrop M. M. (1992) *Complexity, the emerging science at the edge of order and chaos*. New York: Simon & Schuster Paperback; ISBN 0-671-76789-1

(Waring, 1996) Alan Waring (1996) Practical System Thinking. *International Thomson Business Press,* ISBN 0-412-71750-6.

(Department of Transport, 1989) GREAT BRITAIN. DEPARTMENT OF TRANSPORT (1989), *Investigation into the Clapham Junction Railway Accident (Anthony Hidden QC)*. London. Her Majesty's Stationery Office, ISBN 0 10 1082029.